Becoming a Writing Researcher

Becoming a Writing Researcher

Ann Blakeslee and Cathy Fleischer
Department of English Language and Literature
Eastern Michigan University

2007

LAWRENCE ERLBAUM ASSOCIATES, PUBLISHERS
Mahwah, New Jersey   London

Copyright 2007 by Lawrence Erlbaum Associates

Lawrence Erlbaum Associates, Inc., Publishers
10 Industrial Avenue
Mahwah, NJ 07430

Cover design by Tomai Maridou

Library of Congress Cataloging-in-Publication Data

Blakeslee, Ann M.
  Becoming a writing researcher / Ann Blakeslee, Cathy Fleischer.
      p. cm.
  Includes bibliographical references and index.
  ISBN 0-8058-3996-8 (cloth : alk. paper)
  ISBN 0-8058-3997-6 (pbk. : alk. paper)
  1. Technical writing. I. Fleischer, Cathy. II. Title.
  T11.B55 2007
  808'.0666—dc22   2006034032
                                                              CIP

Books published by Lawrence Erlbaum Associates are printed on acid-free paper, and their bindings are chosen for strength and durability.

Printed in the United States of America
10 9 8 7 6 5 4 3 2 1

This book is dedicated to the many graduate students whom we have had the pleasure of teaching and advising over the years and who have contributed so much to our thinking about and perspectives on research and pedagogy.

Contents

Preface

The idea for *Becoming a Writing Researcher* came from our experiences, over several years, teaching and mentoring graduate students and carrying out and reflecting on our own research. These experiences led us to write a text that would foreground the experiences of the researcher (particularly the novice researcher) and that would emphasize the process of doing research, with its recursive and highly contingent nature. We especially wanted to address the experiences of students and new researchers as they begin to undertake research; we wanted to portray research in a realistic manner, with all of its human and messy dimensions.

Our aim with this text is to describe and explain various research methods (our focus is on qualitative inquiry) and to teach students about and socialize them into the experience and practice of carrying out research, be it for scholarly purposes or for more practical or professional reasons.[1] In short, our text responds to two central questions for new writing researchers: What does it mean to be a researcher? And, how do I become a researcher? It is written primarily with these novice researchers in mind (we envision it being adopted as a core text in introductory research methods classes in master's or PhD programs in rhetoric and composition, the teaching of writing, English education, and/or technical and professional communication, all areas in which one or both of us teach[2]). In short, *Becoming a Writing Researcher* introduces students, in a gradual and contextualized manner, to approaches and strategies for carrying out qualitative research, and to the various concerns and issues they will face as writing researchers. It emphasizes the process and experience of doing qualitative research, and focuses on the identities, ethos, and experiences of researchers—especially new researchers—in planning, carrying out, and writing up their research.

---

[1]We acknowledge and address in this text how professionals in writing (whether university professors, professional writers, administrators, or teachers) research in a variety of settings and for a variety of reasons. Not all of our students become academics and scholars in composition and rhetoric. We therefore address and provide examples of a variety of types of research (e.g., teacher research, research for projects in business and industry, academic research).

[2]Blakeslee specializes in rhetoric and technical and professional communication, and she has also taught classes in the teaching of writing. Fleischer specializes in English education, composition, and the teaching of writing.

Because of its focus, this text is organized somewhat differently. Instead of moving students through descriptive accounts of methods and approaches to research, we instead move them through each of the stages of doing research—from considering their identities as new researchers, to locating and articulating a research question, to planning their research, to selecting and using tools to investigate their question, to interpreting and making sense of their findings, to writing up their findings. We also challenge the apparent linearity of our own organizational scheme by continually emphasizing the recursive and dynamic nature of all research. Further, we attempt to make all of these stages in the research process accessible to novice researchers by providing numerous examples and prompts aimed at helping them as they progress through each stage.

Finally, we also structure each chapter in a parallel manner. Each chapter begins with a discussion of *general issues*, in which we address basic concerns about the portion of the research process discussed in the chapter, followed by a discussion of the various physical, conceptual, and strategic *tools* that allow a beginning researcher to carry out that part of the research process. Each chapter also contains a section on *lenses,* where we address the personal and theoretical perspectives and biases that influence that stage of the research process, and a section on *ethics*, where we explore the ethical issues associated with the stage of the process being addressed. The last two sections are especially unique in that we show how personal and theoretical lenses are significant at all stages of the research process and not only in the interpretation stage, and that ethical issues impact research from the very start of the process.

Acknowledgments

Of course, any undertaking such as this is supported by numerous individuals. Over the years, we have had a great deal of assistance in developing this text from both our students and our colleagues. Many master's students in the Written Communication program at Eastern Michigan University (EMU) have used parts or the whole of this text and provided us with very valuable feedback. We are especially grateful to the following students, who have also graciously allowed us to use their work and ideas as examples in the text: M. Diane Benton, Jennifer Buehler, Julie Caldwell, Christian Casper, Amber Clark, Elizabeth Donoghue Colvin, Rhonda Copeland, Deb Fitzgerald, Karen Hoffman, Karen Reed-Nordwall, Erica Pilon, Diane Pons, Dawn Putnam, Brian Roddy, Erin Snoddy Moulton, Lisa Tallman, and Mary Lou Wolfe. Rhonda Copeland also served as our editor, catching both small and more significant issues we had somehow overlooked. She was invaluable, especially toward the end of the process. Liane Robertson assisted us with creating the index for the book. We are also grateful to our colleagues, both at EMU and elsewhere, who used or read portions of the text and/or who provided us with anecdotes to include: Nancy Allen, Todd DeStigter, Heidi Estrem, Bob Fecho, Linda Adler-Kassner, Steve Krause, Barbara Mirel, Frances Ranney, Rebecca Sipe, Dorothy Winsor, and Bob Yagelski. And, finally, we wish to thank our spouses and children—Andy, Seth, and Jesse Buchsbaum and Kim and Sarah Beneteau—whose support and love always sustain us.

1

Me, A Writing Researcher?

Once, during a student presentation in one of our graduate classes, the presenter asked everyone in the class to write down as many nouns and adjectives as we could think of that best described our roles and lives. The words people wrote ranged from "writer" to "student" to "mother" to "teacher" to "runner" to "Native American." Interestingly, not one person listed "researcher," despite the fact that all of these students were working on their master's degrees and many were also starting work on their culminating master's projects. For the graduate students in our program, many of whom are already working professionals (both teachers and professional writers), this failure to call themselves a researcher probably is not that uncommon.

When we looked back on our own graduate experience, we realized that coming to call and think of ourselves as "researchers" also took some time. It was not something that occurred automatically when we became graduate students. For us, and for many others we know in our disciplines, the identity of researcher was at first a foreign one, which for some of us even seemed at odds with our other professional identities (as teachers, writers, and other kinds of professionals). Research was something that other people did—serious scholars who devoted years to their work and who sometimes even secured big grant money to pursue their research. For us, on the other hand, research was what we needed to do to write papers for our courses and to write our theses and dissertations. It was necessary for earning our degrees and, although we both enjoyed it, we initially did not think of ourselves as researchers, at least in the same sense as the scholars whose work we read.

However, as we made progress toward our degrees, our idea of what it meant to be a researcher changed: We started thinking of ourselves as *producers* of research rather than just *consumers* of it. Along the way, we became excited about the idea of doing research, and we began to realize that we could contribute something meaningful to our disciplines. We realized that rather than losing our other professional identities, we could use our identities as writing researchers to enhance our professional work. Our identities as teachers, writers, and researchers could be blended. We also realized that "doing research" was really not all that foreign to us—we actually had been conducting research all our lives.

We Are All Researchers Already

Consider your own thoughts about and experiences with research. Think about all the ways you already are a researcher, focusing for now on your personal life. In all likelihood, you have carried out research for personal reasons many times. You might have done research to buy a car, plan a significant event (e.g., a wedding or graduation), move to a new city, find a good graduate program, or determine the best way to finance your education. We have provided a prompt (Prompt 1: Researching for Life Purposes) to help get you thinking about the personal research you have done. Throughout the text, there are a number of such prompts that, through asking you to write in response to a question or issue, will help you develop your thinking about research. You can respond to these prompts on your own, or you may be asked to complete them as part of a class assignment. We also periodically provide examples or scenarios to show how we or others have experienced certain aspects of research. At this point, we ask you to try the first prompt and read over the examples of research we have done in response to personal needs.

Prompt 1: *Researching for Life Purposes.*

Reflect on one recent experience you've had with research in your own life. What was the experience? What were you trying to determine and what did you do to determine it? In other words, what kind of research did you do? Was your research productive and successful? Was it frustrating? Why? Write up your response to this prompt as a scenario, like those we present.

When you look over the scenarios—yours in response to the prompt, as well as ours—note the commonalities. The motivation for our research arose from specific personal needs: the need to learn about a disease and its impact on a child's life and the need to learn more about the housing market in order to find the right house. To obtain information, we took the following simple and straightforward steps, which you also may have observed in the research you wrote about:

- Searching our memories for our existing knowledge of the subject.
- Talking to other people about what they knew about the subject.
- Consulting outside sources (books, magazines, the Web).
- Analyzing and comparing our sources and sorting through contradictory information.
- Arriving at some kind of new understandings that helped us take action.

Sidebar 1: *Researching for Life*

Cathy: When I wrote this chapter, my husband and I were in the midst of a personally important research project involving our then 8-year-old son. After watching him experience what we called "space-outs" for a few months, we began—with our doctor's urging—to carefully watch how often he had them and if there were any triggering circumstances. After we noted a marked increase in the number of incidents, we brought our information in to the doctor, who suggested a possible diagnosis of absence epilepsy. Our initial shock at this diagnosis was based on our limited understanding of the term and the associations that epilepsy brought to mind. We then began to read books, analyze Web sites, and talk to friends who knew children with similar diagnoses. Armed with a lot of information, we were able to talk to the doctors with some well-formed questions in mind and with an awareness of the various treatments, side effects, and long-term prognoses. As our son started a regimen of medication—a regimen we learned was very difficult to regulate—we kept a chart in which we noted each day how many seizures he had, how he slept that night, what kinds of food he ate, and any other factors that might be involved. Each time we saw the doctor, we referred to this chart and pooled our information with the doctor's expertise in order to make sound treatment decisions.

Ann: While Cathy was writing the initial drafts of this chapter, my husband and I were moving into our new house. Our move was the culmination of a long process that lasted more than a year. We knew that we needed a larger house, so we began to peruse the real estate section of our local newspaper and to attend open houses. We also started talking to friends about different neighborhoods, and we began talking to banks about mortgages and interest rates. We started out considering homes in a fairly large geographical area because we wanted to reduce my husband's commute while keeping my commute within reason. We visited numerous towns in the area, and we also picked up real estate books for those towns. We looked on the realestate.com Web site, and we compared property taxes for different locales. Because we were interested in new construction, we also asked people about builders, and we consulted people we knew in the construction and real estate fields. We collected numerous architectural plans as well. Finally, we located the house we purchased through a for-sale-by-owner advertisement in the Sunday paper. The ad listed a URL, so we viewed the house initially on the Web and then went to see it. We ended up buying the house, which led in turn to a good deal of additional research—we had to determine a price for the house we were selling, we had to hire painters for the new house, and we had to select a mortgage company. All in all, we talked to a lot of different people, did quite a few Web searches, and purchased numerous home magazines. We have continued doing similar kinds of research as we have made improvements to the house.

In addition to consulting secondary sources and talking to people who knew more about the subject than we did, we also undertook some of our own firsthand research: in the first case, observing a child carefully to ascertain when the seizures occurred and how the drugs affected him; and, in the second case, timing the drive to work from different areas to make sure that the commute would be manageable. We also developed ways to organize the information we collected. Cathy developed a chart to record her son's responses to the medication, and Ann created paper and computer files to store information. In both cases, the research was not only ongoing (taking place over a number of months), but it also had an impact on others (ourselves, our spouses, and, in Cathy's case, her son and his doctor). In addition, we both broadened our understanding of our respective problems and needs, and we did so with the expectation that others might ask us to share our newly acquired knowledge—which they did. We suspect that your research stories reveal similar strategies and outcomes. You can use the second prompt we have included to draw comparisons.

Prompt 2: *Identifying Research Strategies.*

Review your own research stories and draw comparisons to ours. What strategies did you use that were similar? What strategies did you use that were different? What did you do with the information you collected—how did you sort through and organize it? What outcomes did you achieve and what actions did you take as a result of your research? Did anyone else benefit from your research? If so, who and in what ways?

In addition to any personal research, you probably have also done a good deal of research in your academic/professional life. For example, as a student you have no doubt researched and written papers at just about every stage of your education. If you have worked already as a professional, then you likely have also done research for your job. For example, if you are a teacher, then you probably have researched your lesson plans, and maybe even used research to address problems or questions (e.g., why your students have trouble revising). If you are a professional writer, then possibly you have researched software applications or studied company documents to determine their tone and style.

Prompt 3: *Performing Research in Your Field.*

Think of two occasions on which you conducted research for academic or professional purposes. Address what motivated you to carry out the research (e.g., a class assignment, a need or problem in your workplace, etc.). Also address the strategies you used and indicate which ones seemed the most and least productive for you, and why. Finally, address what your research contributed to your work or studies or how it helped you.

These examples make a point: You already know a great deal about research from the personal, professional, and academic research you have done. You have certain steps you follow that help you do research: steps for gathering information, analyzing it, and taking action as a result of it. And, most of all, you know that doing research matters: It helps you make decisions and achieve specified results. From our vantage point, this understanding that *research matters*, personally and professionally, is essential in the process of coming to think like a researcher.

So why is coming to think like a researcher important? What does it accomplish? On the one hand, it enhances our lives as professionals. Jennifer Buehler, who taught high school English in Plymouth, Michigan, explains it this way:

> Thinking like a researcher has helped me to feel more engaged as a teacher. I'm now interested in much more than the effectiveness of a simple lesson plan or assignment. Conducting research has caused me to see the students as complex people and the classroom as a complex culture. I look now with the eyes of an ethnographer, and, as a result, what I see is richer and far more challenging to process. I believe that because of my research, I bring more thoughtfulness and insight to the classroom—reflection is now at the forefront of all I do. I am more committed to remaining a teacher than I was before becoming a researcher ... [It's] added depth and meaning to my work that I would not have found otherwise. (personal communication)

Karen Reed-Nordwall, who taught middle school English in Livonia, Michigan, explains:

> Research is definitely not separate from teaching. Research is teaching, just like revising is writing. I can't believe how much my teaching has improved since I've realized this. Researching has brought everything I do to a hyperconscious level. It's also made me realize that I have a voice. I see these kids every day and know where they are at better than any other researcher or textbook maker. Doing research, studying what my students need to learn and practice, made me realize that I don't have to always look outside my own classroom for answers. I have learned to trust what I see on my paper, in student conversations, and in my classroom to help me teach my students. (personal communication, 23 March 2003)

Amber Clark, a technical communicator who worked at Tec-Ed Inc. in Ann Arbor when we were writing this book, responds:

> My company has a commitment to ongoing professional development through reading and research. Our reading keeps us current with new developments and ideas in the field of technical communication and usability, while our research enables us to contribute to that general body of knowledge. Most of our research stems from experiences with specific clients, and usually involves new methods or design ideas. We find that it's important to have "back burner" research projects going along with

our client projects so that we can stay on the cutting edge of the technical communication field. (personal communication, 15 April 2003)

Finally, Elizabeth Donoghue-Colvin, a technical communicator who was working at Blue Cross Blue Shield of Southeast Michigan, reflects on how research influences her professional life:

> I notice things about the process of writing that I might not have paid much attention to before. I've also begun to organize my thinking about everyday observations in ways I would not have before I became involved in research. … The point is that, because I'm involved in research, I'm taking notice of the things that are going on around me at work more than I otherwise would have. I'm thinking about the process of writing. (personal communication, 24 February 2003)

One thing that we find interesting in these scenarios is that the research these individuals do in their professional roles shares the same primary goals as the research all of us do for personal reasons: to help us become better informed and more knowledgeable and to help us make decisions and take action. All of these individuals agree that research is an important part of what they do at work; it's a central part of their day-to-day professional activity. The direct and firsthand information they gain by doing research, whether it comes from the students they teach or from the users of their written documents, is invaluable to them. And perhaps most significantly, a few years ago, these researchers were graduate students in a master's program, perhaps just like you. In a very short time, they have each come into their own as researchers and they have come to view research as an integral part of what they do professionally.

Becoming a Researcher: What Else is Involved?

As we read the responses of these professionals, it is hard for us to believe that just a short time ago they were beginning their culminating master's projects with numerous fears about their abilities as researchers. We have asked ourselves (and them) how it is that they managed to gain the knowledge and confidence to now think of themselves as researchers. What did they do to be able to integrate research so seamlessly into their professional lives? What were the challenges and successes they encountered as they did so? As we have thought about our own growth as researchers, and watched these and many other students develop into thoughtful and reflective researchers, we have been able to identify at least three factors that contributed to their success. First, there is the recognition of the important role research plays in both our personal and our professional lives, which we have been addressing throughout this chapter. Second, there is the realization that research is something we all actually know a good deal about already. Third, there is an interest in learning what is involved in doing research and how research is done.

Recognizing the Important Role Research Plays in Our Lives

Ann Berthoff, an important figure in both composition studies and the teacher research movement, once suggested that we pronounce "'research' the way the southerners do: REsearch." She made this suggestion, she said, because, "REsearch, like REcognition, is a REflexive act. It means looking—and looking again. This kind of REsearch would not mean going out after new 'data' but rather REconsidering what is at hand" (Berthoff, 1987, p. 30). Berthoff's suggestion carries with it an important consideration. If we think about research as searching again, as looking carefully and reflectively at what is in front of us, then we can demystify the research process. Research, in other words, does not always involve complex, multitiered studies in which the researcher enters into unfamiliar settings to discover new "truths." Equally valuable are research studies that are focused instead on settings right in front of us, places where we can reflect carefully and methodically on issues of concern in our own work or for our own personal or professional curiosity and growth. The research we carry out need not be Research with a capital "R"; it can be research with a lower case "r"—small, manageable, and reliant on the work of others. In short, this focus on research as a reflective activity is key, especially for those of us who see research as a part of our everyday lives and professional commitments.

For this kind of research to be successful, it is vital to begin with a true wondering, that is, with a question about which you are indeed curious, whose answer might create new understandings for you in either your immediate setting or in your own more general professional growth. This approach to research is a natural extension of the personal research we asked you to reflect on earlier. Just as personal research becomes meaningful as we work to find answers to questions that have an immediate impact on our lives, so too does professional research when we discover questions that might truly impact our professional growth and development, and even our fields.

Realizing That Research Is Something We Already Know a Good Deal About

As you begin the process of becoming a researcher, we believe it's vital to consider first what you already know, both about the subject you are researching and about the kinds of research you are considering. Inevitably, new researchers discover that they, indeed, do know quite a bit already. You know about the context of your own work, which is knowledge that will be key to developing a research question. And, as people who have taken numerous classes, you also know quite a bit about your field and about its collective knowledge. You probably have read a lot in the field, and you are already aware of many issues that interest researchers and practitioners in it. As a result of your reading, you probably also know about the methodologies used by researchers in your field, and you most likely have developed a feel for those methodologies you would like to use for your own contexts and questions. The recognition that you hold this knowledge will hopefully help you realize that you are not a beginner. You already have a strong foundation. This section's prompt will help you identify what you already know, both about a subject that interests you and about research more generally. You can also use it to begin identifying what you still need and want to learn.

Prompt 4: *Identifying What You Already Know.*

Take a minute to respond to the following questions and then share your responses with your classmates:

Questions About the Subject of Your Research:

- What are some of the things you already know about the subject you're interested in researching? Make a list.
- What are some of the questions you have about the subject you're interested in researching? Again, write all of them down.
- What have you read that addresses your subject? What have the sources you've read said about the subject?

Questions About Doing Research:

- What do you already know about doing research?
- What kind of experience do you have with carrying out research?
- What do you believe makes you qualified to do research on writing?

Learning What Is Involved in Doing Research

One fear that many students voice (and that we, too, felt at one time) is a perceived inadequacy in regard to how to do research—that is, how to design a research study, how to go about gathering information, how to sift through the information to make meaning of it, and how to write up the findings so that others find what we say interesting and meaningful. Part of that fear may rest in the image you have of who a researcher is and what a researcher does. When we asked our own students what they thought of when they thought of researchers, many conjured up images of men in white lab coats (and it usually was men) gleaned from portrayals of research in movies and textbooks. Many also thought initially of more experimental and quantitative types of research. Such limited conceptions, we would contend, make research seem even further removed from our everyday lives and work.

An alternative conception, and one that is more consistent with how we portray research in this text, relies heavily on the relationships between researchers and their individual informants: students, other writers, and so forth. Writing research tends to be more qualitative, but no matter what kind of research is done or what methodological tools are employed, the decision to do that type of research should be a deliberate and thoughtful one. These decisions depend on reflection as well as on personal beliefs, such as those about what research means and how knowledge is created.[1] For example, the preference both of us have for qualitative research has a great deal to do with our views of the world and how knowledge exists in the world. As social constructionists, we believe that meaning is a concept that is created among participants within the constraints of particular social contexts. In other words, we do not believe that there is a "truth" out there that we as researchers can discover; rather, we believe that meaning is always contextual, ever-changing, and exceedingly complex. This belief leads us to see research as an attempt to get close to some kind of meaning in particular contexts; it leads us to work closely with a small number of people to try to understand their stances toward a subject and to try to make sense of these understandings in a deep, contextual way. Our goal is not to generalize from these situations to the rest of the world; rather, it is to take a particular research case and place it up against many research cases in order to begin seeing similarities and contradictions at work.

While admitting that our own biases have often rested with the qualitative side of the research debate in writing, we also recognize that it does little good, and is not very productive professionally, when either side levels accusations against, or relies on, stereotypes to describe the other. Further, the stereotypes these kinds of debates tend to propagate usually do not hold up. Most scientists we know, for example, do not fit that stereotypical image conjured up by our memories of what

---

[1]We want to acknowledge here the debates that have sometimes occurred in composition studies regarding the relative merits of qualitative and quantitative approaches to research. Qualitative researchers sometimes have claimed that quantitative researchers are too focused on numbers and too concerned about tightly controlled experiments, so much so that they miss or completely overlook context. Quantitative researchers, on the other hand, claim that qualitative researchers are too relativistic, too inclined to generalize from single cases, and unable to see the larger picture. These researchers question the ability of qualitative researchers to prove anything, especially with an "n of 1."

we saw on television or read in our elementary and middle school science text-books. And most qualitative researchers approach their work with great attention, care, and rigor, which we hope you will learn to do as you read this book and carry out your own research. All kinds of research are valuable. What matters are the questions researchers ask, the purposes they have for the research, the circum-stances surrounding the research, and the kind of impact researchers hope to make. All kinds of research play important roles in the creation of knowledge and in the advancement of understanding in our field. (For further discussion of these issues and their potential impact on our field, see Charney, 1996, 1998, and Barton, 2000.)

Certain methodological tools, however, are natural extensions of the belief sys-tems that underlie particular kinds of research. Because qualitative researchers, for example, believe in the importance of context, a key tool in this kind of research is observation. Because we also believe in the importance of listening to and preserv-ing the voices of others, another key tool is interviewing. Finally, because we value the materials our research participants create and use, a third key tool is artifact analysis. These tools (and others) are learnable. In fact, we have discovered, both through our own experiences and through the experiences of our students, that learning about these tools—developing a fuller understanding of what researchers do to undertake their research—helps with gaining the confidence you need to begin calling yourself a researcher.

Another key to understanding what is involved in doing research, and then gain-ing confidence as a researcher, is realizing that learning to do research takes prac-tice. It's something you can get better at over time. As you carry out different kinds of studies, you learn new strategies for doing research, and you may even change your beliefs about what constitutes good research. You realize that you grow and change continually as a researcher, and you come to understand that there is no sin-gle correct way to do research. Rather, you learn to reflect on and think critically about what you do in your research, and you realize that each successive study you do gives you the opportunity to do it better.

HOW YOU WILL ENCOUNTER RESEARCH IN THIS TEXT

Our goals for this text include helping you learn about the experience and practice of qualitative research and helping you begin to develop an identity as a writing re-searcher. We hope this book will help you become comfortable and confident with research so that you can integrate your research seamlessly into your professional life. The chapters that follow emphasize the process and experience of doing re-search and ask you to reflect continually on your identity, your ethos, and your ex-periences as you plan and carry out a qualitative research project.

The text moves you through each of the major stages of research. However, bear in mind that real research is not linear and sequential, as is suggested by the delineation of these chapters. Real research, rather, is recursive in nature and is a process, much like writing itself. Still, for the purposes of the text, we devote at least one chapter to each of the four main stages in the research process:

1. Finding and articulating a research question (chap. 2).
2. Planning your research (chap. 3) and selecting methods and tools to investigate your question (chaps. 4 and 5).
3. Sifting through and making sense of your research findings (chap. 6).
4. Writing up and presenting or publishing your findings (chap. 7).

Each chapter has the same basic structure. The chapters begin with a discussion of general issues surrounding that stage of the research process (what that stage involves, how you can get started in that stage, and how other researchers have successfully navigated that stage). These general discussions, we hope, will both introduce you to the particular stage and help you begin considering generally how your own individual project might take shape. We then get more specific about that stage of the research process by identifying a range of tools and strategies for it that we and other researchers have used with some success. Written to encourage a hands-on approach, these sections integrate a discussion of tools and strategies with specific activities and prompts that you can apply to your own research project. We suggest that you use the prompts in a way that makes sense to and helps you as a beginning researcher. For example, you may want to start a research journal in which you record your responses and create a record of your developing thoughts about research. You may also want to select among the prompts, doing only those that seem meaningful to you. Your instructor may also ask you to respond to particular prompts.

Sidebar 2: *Keeping a Research Journal*

Many researchers are committed to keeping a research journal in which they record their thoughts and ideas throughout the research process. Researchers list their multiple versions of their research questions, their responses to readings, their frustrations, their initial thoughts on methodologies, their preliminary and developing interpretations of their findings, connections they begin seeing, and so forth. We encourage you, as you work your way through this text, to try out a research journal and to use it to record your developing thoughts and ideas about your research. You can also record responses to the prompts in your journal. We have written the prompts to guide you through the various stages of the research process and to help you anticipate and address the concerns and questions you may have as you prepare for and then carry out your research.

Each chapter also explores the role of personal and theoretical lenses for that stage of the process. We ask you to consider how your own experiences and viewpoints, and those of others, can affect your work as researchers. Our belief is that considering these lenses is essential for all researchers, but especially for qualitative researchers. Each of us brings cultural, ethnic, gender, class, theoretical, and occupational biases to our work, which create certain lenses through which we view what we observe and hear. We contend that researchers cannot ignore or suddenly eliminate such biases. Rather, we need to be very aware of these stances and awake to their implications for our research. We also need to be self-conscious about how we represent them in our writing. In this part of each chapter, we address how our personal and professional lenses may influence that stage of the research process, and we offer suggestions for both embracing and countering these biases, depending on the situation and need.

Finally, we end each chapter by identifying and discussing some of the many ethical issues that attend that stage of the research process. As committed qualitative researchers, we are concerned continually with the issue of ethics: how we treat and represent the people and sites that are at the heart of our research, and how we tell the story of our research. Ethical questions and concerns are at the heart of any research project—questions, for example, about whose research it is, who benefits from it, and so on. And although these questions may have no easy answers, they raise our awareness of the complexity of doing research, especially that kind of research that involves real people in real situations. Throughout this text, therefore, we devote a section of each chapter to addressing the ethical implications of that stage of the research process. We hope that doing so encourages you to think continually about the ethical aspects of everything you do as a researcher.

Learning to become a researcher, we believe, is a journey that, like any other, is filled with both excitement and challenges. Our hope is that this book will guide you in that journey on both practical and theoretical levels. We hope it will help you complete the task immediately at hand—completing a required methods course or writing a thesis—but that it will also help you begin to think of yourself as a researcher in all of the academic, professional, and personal endeavors you undertake. We want you to think about how you might incorporate a research stance into your life, whether as an academic for whom research is an integral part of your everyday work, as a professional for whom research functions to inform and improve your everyday work, or as a concerned and informed individual for whom research functions to enrich your everyday life. Thinking like a researcher hopefully will become a way of life for you—as it has for us—that helps you reflect on and continually improve what you are doing, both personally and professionally.

2

What's Your Question?

In the first chapter, we suggested that you begin your journey as a researcher by considering what you already know. This chapter presents strategies to help you expand on that knowledge and use it to formulate a research question. When we talk about a research question, we mean the question that expresses what you wish to learn through your research. Your research question is what will guide and direct your research.

Although it sounds like a simple thing to do, articulating a good research question can be a challenge. Some scholars contend that research studies are only as good as the research questions behind them, which can put a great deal of pressure on you as a new researcher: Concerned that this initial stage of identifying a question might affect your whole research process, you may, for example, end up wanting to develop the perfect question from the outset. You might also be plagued with doubts about the value of your question: You may worry about whether anyone else will even care about it. Another concern students often express is that they have not yet read or worked enough in the field to generate a good question. And then there's the issue of scope and whether a question is too small or too big. Finally, you may be concerned with whether your question is interesting enough to sustain your attention through an entire research project. Any of these concerns can make the task of formulating a research question seem daunting.

For that reason, we would like to suggest a mind-set for you to adopt as you begin thinking about your research question. We encourage you, first, to think of research as an exploration that is guided by some purpose—by that desire to accomplish the kind of personal or professional goal that was discussed in chapter 1. Research questions are intended to express that goal. Further, remember that research is about *inquiring*: You carry out research to learn something new, to answer a question, to solve a problem, or to fill a gap. Research is also a *process* that is normally recursive. In other words, you go back and forth between the various activities entailed in research, which suggests that the things you do, including the questions you define, usually are not set in stone. In fact, sometimes you may not even settle on a research question until you have already carried out some of your research. Note one researcher's experience in this regard:

> I often do a fair amount of data gathering before I settle on a question. I start out with a general sort of area I'm interested in (say, writing in an engineering center), but I don't actually know how to ask a good question until I see what's happening. How can I know what will be interesting to look at further until I've looked some? (Dorothy Winsor, personal communication, 25 January 2003)

This chapter discusses a number of strategies, including gathering preliminary data (like Winsor does), that we hope will help you articulate a good research question.

Prompt 1: *Identifying Personal Traits.*

Take a moment to think about, and jot down, the personal qualities you have that will help you to be a good researcher. You might, for example, be conscientious, detail-oriented, careful, inquisitive, caring, or any number of things that will help you as a researcher. Because the early stages of the research process are sometimes challenging, it can be helpful to reflect on the qualities you already have that will make you a capable researcher. So generate as long of a list as you can and refer to it often.

GENERAL ISSUES IN DEVELOPING A RESEARCH QUESTION

The following general issues are discussed in this section:

- Assessing your interests and experiences
- Remaining open to new ideas
- Acquiring the habits of researchers
- Liking your research question
- Assessing your investment in your question
- Understanding the dynamic and recursive nature of research

Assessing Your Interests and Experiences

If you are reading this text as part of a class, you will likely find a good deal of variety in your classmates' research interests and in how well defined they are. Some of you will have a pretty clear idea of your interests already, whereas others will have only a general sense of them. Some will not be sure at all yet. The one thing you'll have in common, however, is that all of you *have* experiences and interests that will impact what you finally decide to research. In fact, before you read any further, you may wish to complete the prompt for this section, which will help you inventory your professional and personal interests.

The lists you generated in response to the questions in the prompt should help you begin thinking about what you would like to research. However, you should

---

Prompt 2: *Inventorying Your Personal and Professional Interests.*

Jot down responses to the following questions. Write as many responses as you like to each question. Your goal at this point is simply to identify interests, not to narrow them:

- What issues interest or concern you the most in your field?
- What do you want to learn more about?
- Of the reading you have done in your courses or on your own, what has been most interesting to you? What are some topics you have read about that you find meaningful?
- As a practitioner in your field (if you are one already), what's been the most interesting to you? What contradictions have you observed between what you have read and learned in your coursework and what you have experienced as a professional?
- If you are not a practitioner yet, what observations or insights have you gained in discussions with your classmates and instructors? Have any of your classmates or instructors identified issues in your field that interest you and that seem to warrant further research?

Now that you have responded to each of the previous questions, go back and give some thought to these questions:

- Why do each of the topics you listed interest you?
- Why do you want to learn more about the topics you listed? What do you want to learn about them?
- Why do the topics you identified from your reading and other sources hold meaning for you?

You may find that you need some time to reflect on these questions. That's fine. Give your ideas time to develop for a few days before you jot down anything. Or jot down some initial ideas now, but make it a point to revisit the questions. Because research is recursive, you can, and really should, revisit these questions at various points in your work. Pay attention to how your responses change as your knowledge of the topics changes.

---

also consider your own experiences. In many cases, the topics you end up researching professionally will hold special meaning for you personally as well—they may be tied to your background or personal experience. For example, one of our colleagues, Becky Sipe, carried out research on spelling for several years. Becky's interest in spelling is as much personal as professional. Although professionally she works in the area of English education, training future middle and secondary English teachers, personally she has struggled with both her own and her son's challenges with spelling. In another example, a colleague at a nearby university, Frances Ranney, has been carrying out research on legal discourse since she was a graduate student. Before returning to school for her doctorate, Frances worked for several years as a legal assistant. She brought to her research, therefore, extensive experience with the discourse she chose to study.

Remaining Open to New Ideas and Acquiring the Habits of Researchers

So the areas you research and the questions you pose for your research are often connected to your personal and professional interests and experiences. They have a history. Even if you are delving into a new area, you frequently will possess some special interest in that area—for example, you will want to know more about it because it is starting to impact you in some way. Of course, when you are just beginning this process, and especially if you are not yet sure of what does interest you, it helps to consider a range of possibilities. It also helps to remain open to new or unexpected ideas. Again, research is an exploratory process; its root, "search," truly captures its essence. Researchers wonder about, experience, look for, and inquire, and they do these things continually. They develop habits. For example, when they read, they jot down ideas and questions; when they observe or experience a situation, they note those things that strike them in some way; when they hear an interesting lecture, they write down their responses and formulate questions for the speaker. As you begin to think of yourself as a researcher, you will acquire these habits. In fact, before you know it, you may even start doing some of these things intuitively.

Liking Your Research Question

It may go without saying at this point, but you should also be sure that you like your research question. For example, if you really like technology, then you should give serious thought to formulating a research question that incorporates something about technology. In fact, if you have any doubts about your interest in a topic, then you might want to consider other topics before formulating a question. Because you will likely make a long-term commitment to your question, you do not want it to become boring after a short amount of time.

The following are some reasonable questions to ask yourself at this stage: "What do I really care about?" "What am I passionate about?" Sometimes the answers to these questions are obvious, but at other times you might need to consider such questions carefully. It's good, as well, to think about what's behind your interest in a topic. For example, some of our students have defined a research question based on something that was important in their workplaces. In many cases, this has turned out fine; however, in a few cases, events occurred that made it problematic—their priorities in the workplace shifted, and/or the students ended up leaving that workplace. These students either were unable to complete their projects, or they had a great deal of difficulty completing them. We offer this simply as a caution for those who may define a research question that is connected to their workplace. You might well want to consider whether and how the question could be adapted should your situation change.

Assessing Your Investment in Your Question

We need to offer another caution as well. Although it helps a great deal to love your research topic, it is also possible to love it too much. In other words, as a researcher you sometimes have to guard against blind spots that may arise from the personal or professional investment you have in a topic. We address this in greater detail in the "lenses" section and in the next chapter, but for now we mention it as a caution. It is helpful to assess and articulate why you are interested in a topic, what your investment in the topic is, and what is at stake for you in researching the topic. This is not to say that you should shy away from what you really love. Quite the contrary. That's what will motivate you the most. However, you should be aware of your stance and positioning in relation to your topic: Consider, carefully, the biases you might hold and the judgments you may have formed already because of your interest in your topic. Sometimes, reflecting on your stance may be enough. Once you're aware of your biases, you may be able to pursue your research in an objective manner. However, if a topic evokes a very strong emotional response from you, whether it is anger, compassion, or any other strong emotion, then you should think about how (or whether) you can carry out your research without those emotions clouding your judgment. Having too strong an investment in your topic can limit or shut down the discovery process. We address these concerns and their implications further in subsequent chapters.

Understanding the Dynamic and Recursive Nature of Research

Something else you should keep in mind at this stage of the research process is that your interests and research questions are dynamic. In fact, they very likely will change as you deepen your understanding of the area. Sometimes just realizing this takes some pressure off. You realize that you do not need to define the perfect research question right off the bat. You also realize that you probably do not even know enough at this stage to do that. The research process is gradual and ongoing.

You do not complete one stage and simply move on to the next in a linear progression. You learn constantly when you carry out research, and whenever you learn something new your thinking about or approach to a topic may change. And, as it does, you may end up reformulating your research question. What is essential is giving yourself permission to do that.

TOOLS FOR DISCOVERING, ARTICULATING, AND NARROWING QUESTIONS

So what's the best way to go about discovering the area you wish to research and formulating your research question? Fortunately, there are a number of tools you can use to help you discover, articulate, and even narrow your question. We present several of these tools in this section:

* Taking personal and professional inventories
* Reading
* Observing and noticing
* Talking with others
* Writing in a journal or log
* Looking for recurrent themes
* Gathering preliminary data
* Broadening and narrowing
* Phrasing the question

Taking Personal and Professional Inventories

Probably the best place to start is with the inventories of your personal and professional interests that we asked you to construct in the previous prompt—your answers to questions about your specific interests and your knowledge of your field. Developing such personal and professional inventories can help you brainstorm and identify some of your possible interests. It can also help you begin identifying those things that do not interest you as much as you originally thought. By externalizing and getting your thoughts down on paper, you can start looking at them differently and you can consider different perspectives on and options for considering them.

Reading

A second tool, one that you will likely use even as you construct your personal and professional inventories, is reading. Reading serves several purposes at this stage of the research process. First, it informs you of what's out there already—and what's been done on and said about a particular topic. Reading gives you a map of your field and of your areas of interest (see chaps. 3 and 4 as well). Reading also stimulates your thinking about topics. When you read, you respond to others' ideas and develop your own ideas. Sometimes your responses may be emotional; at other times, they will just be tied to your own curiosity about a subject. You may find something interesting, agree or disagree with something, or get angry or frustrated in response to something you read. Sometimes, too, you sense a gap in what you are reading, something that is not addressed that you think should be. It may be a gap in the author's argument, or it may be a contradiction between what the author has claimed and what you have observed or experienced personally. Some scholars refer to this perception of a gap in the literature as a *felt difficulty*. You sense that something isn't quite right. And what that something is might be readily apparent to you or it might be something you need to reflect on.

Prompt 3: *Responding to Reading.*

Think of something you have read in your area of interest that you found troubling or lacking, something with which you disagreed, or something that seemed contradictory to your own experience. Identify it and try to articulate the problem or difficulty you had with it.

The gap, or felt difficulty, you perceive in your reading may well become the seed for the research question you formulate. It is useful, then, to pay attention to those things that trouble you as you read, those things that seem incomplete, baffling, or even wrong. Of course, if you are just starting out in an area, you might not trust your instincts. You might wonder how you could possibly know enough about a subject to raise these kinds of concerns. Whereas such feelings are legitimate, your gut reactions to what you read are also legitimate. At the very least, take note of these feelings. Write them down. Finally, you may also end up being inspired by something you read: It may raise new questions for you, or it may evoke other settings where the author's questions also apply. This is an experience that Dorothy Winsor, quoted earlier, addresses:

> Another thing I do is attempt to imitate studies that I have enjoyed reading. I think, "I could do that, only using these participants over here who are slightly different." That's what happened to me when I read [Latour and Woolgar's] *Laboratory Life*. I thought, "I could do that, only with engineers." And when I heard about the Challenger accident, I thought of J. C. Mathes' work on Three-Mile Island, and I thought, "I bet there are similar questions to be asked." (personal communication, 25 January 2003)

We suggest that you keep a journal as you embark on your reading (see chap. 1 and the section on journals later in this chapter) and consider questions like these each time you read something that interests you:

- What was most interesting for you about this piece?
- What was most troubling or puzzling about it?
- What questions does the piece leave you with?
- How does this piece connect to your own experiences?

You can add other questions as they occur to you. If you are visually oriented, then you might develop a way to array the ideas you identify in your reading, perhaps through charts, webs, or graphs, or perhaps by displaying other authors' ideas alongside your own. Such displays can also reveal gaps in the literature, because you can see more readily what has and has not been said about an issue. Finally, talk with others about what you are reading: Talking with a faculty member, classmate, or friend can help you sort out ideas and identify those ideas that really interest, strike, or trouble you. Talking about what you are reading can also help you understand the reading better.

Of course, you have to decide what to read, how much to read, and when to stop reading. Although reading can generate questions and new ideas, it also can overwhelm you. There's a lot out there, and in today's technology-rich world, more of it than ever is easily accessible to us. Sometimes it's helpful to set limits so you don't overwhelm yourself. Also, remember that you have plenty of time ahead of you to read (you will be reading during every stage of your research). When you are at this stage of the process, read broadly but also selectively. Don't feel you need to read everything that's out there. Concern yourself primarily at this stage with getting your bearings so you can formulate an effective question. You can always go back to sources later. Just be sure to record the references for them. (We talk more about reading in chaps. 3 and 4.)

Also, if you are not sure where to start with your reading, identify some articles or books that you've read already (e.g., in your classes) and go back and take another look at them. Look at their bibliographies. What other works do they cite? You might make note of those works that seem relevant. You can also ask a faculty member to suggest sources. And you can carry out your own searches by accessing your library's online catalogue and databases (something we also say more about in chap. 3).

Once you have inventoried your personal and professional interests and begun reading, what's next? If you have a pretty good sense of what it is you want to investigate, then your research question may take shape almost immediately. Some of you, however, may need additional assistance. The following sections present several other tools that can help you with this stage of the research process. Our list is not exhaustive, but it should at least give you a good start.

Observing and Noticing

Sometimes it is helpful just to pay attention to your surroundings. Every time either of us teaches a qualitative research methods class, we ask our students, within the first few weeks of class, to select a location and to just observe what's going on in that location. After they observe, we instruct them to record their observations, along with any reactions they had. One of our favorite responses to this assignment came from a student who entitled her observation "Auto-Ethnography." This student sat in her automobile (a truck in this case) and simply attended to what was going on in and around the truck. Other students chose settings closer to their research interests. For example, a middle school teacher decided to observe her students during their reading time. She was amazed at all the things she noticed them doing as they read.

You should be aware that observing and noticing can actually be quite challenging, especially if you are the kind of person who likes to talk and get involved in situations. In fact, you may need to train yourself just to sit quietly and watch. You'll be amazed by how much you are able to observe, especially if you observe a setting that is familiar to you. The practice of making the familiar unfamiliar is one that most qualitative researchers find advantageous.

Observing and noticing can be one-time deliberate activities, or they can also become habits that you acquire as a researcher. In other words, once you get the hang of it, observing and noticing might be something that you just do naturally. They can also become valuable tools as you carry out your research, which is addressed in chapter 4. When you really get engrossed in your research, you will always be looking for things that apply to it. You will carry your question with you and consciously—and sometimes even unconsciously—make note of those things that pertain to it.

Prompt 4: *Observing Your Surroundings.*

This prompt has two options. For the first option, select a situation and observe it for 30 minutes or an hour. Just sit quietly and pay attention. Do not even take notes until after you have completed your observation. At that time, write down what you observed. Record the ordinary as well as the unusual. Be as thorough as you can. Then, either write a description or give an oral report of your observations to your classmates. If you write up your observations, try to do so in enough detail so that someone unfamiliar with the situation can imagine it. (As a variation of this prompt, two or three students in a class could observe the same situation, but separately. The class can then hear and compare the observations, which will likely be very different. This variation, carried out in one of our classes, yielded some surprising and interesting distinctions.)

A second option is to plan to carry out a more extensive observation connected to your research interests. Take a week and observe situations that have some connection to your research interests. Observe these situations systematically on a number of occasions. After you have completed each observation, record it in the manner described earlier. As the week progresses, revisit your notes and try to develop questions from your observations. Report your observations and questions to your classmates.

Talking With Others

In contrast to some of the stereotypes addressed in chapter 1, research is generally not something that is carried out in isolation. In fact, you will likely find that your own research is more productive when you consult with others. Talking with others—advisors, classmates, a partner or spouse, or even a friend—can be especially helpful at this stage of the research process. Sometimes, for example, it helps simply to talk about what you have read or what you are thinking about. Expressing your ideas to others is a good way to externalize your ideas. It helps you view them differently. And the best part of expressing your ideas to others is getting their feedback. Simply posing the question, "What do you think?" can lead others to share all sorts of viewpoints.

Prompt 5: *Talking With Others.*

For this prompt, complete the following tasks:

- Free write a list of possible research questions.
- Pick out one question that interests you and write it on a blank piece of paper. Fill in the following: "I am interested in this question because ... " Also, "I already know these things about this topic ... , but I don't know ... ." (make lists for each of these).
- In a group of four or five people, pass your response to the person to your left. Everyone in your group should respond to what you have written by using the following prompts: "What I find most interesting about your topic, is ... " (list); "Some questions I have about your topic are ... " (list); or "Some resources I know about that would be useful for this topic are ... " (list).
- Keep passing the papers until they are returned to their authors. Each person should have a sheet with responses.
- Read your responses and free write about what you have learned.

Both of us use the strategy of talking with others, and we cannot recommend it enough. If you are a student, then you have the ideal situation. You have your professors, probably an advisor or major professor, and your classmates. You might even talk to professors outside of your program to get their views on an issue. We also recommend that you consider developing a research group with some of your classmates. One of the times Ann taught the research methods class, six of her students formed a group that called themselves the "Six Scholars." The group continued meeting long past the end of the course. They supported each other as they planned and then carried out their research for their master's theses and writing projects. They even started addressing each other as scholars (i.e., Scholar Tallman and Scholar Benton), which suggests that they had begun assuming the identity of researchers. Sharing ideas with others and hearing their responses to your ideas can be a very productive strategy at this stage of, and really throughout, the research process (as both of us attest in the sidebar).

Sidebar 1: *Talking Through Our Questions*

Ann: The evening before I started writing a draft of this chapter, I got together with a friend to talk shop. Each of us had our own objectives for the discussion that reflected the different stages we were at in our research processes. My friend just sent off a proposal for a book to two publishers, and she already had an expression of interest from one of them. She wanted to talk about the next stage, how she should proceed with drafting the first chapter she planned to write. She was especially concerned about how she would do this since it would be the first time she'd be writing for a practitioner audience (as opposed to an academic audience). I, on the other hand, was concerned with getting my friend's feedback on a new research project, which at the time was only vaguely defined. I knew the general issues I wanted to explore, but I wasn't sure yet how I'd define them or what settings I'd select for my inquiries. I was feeling at sea, and what I most needed was the opportunity to put my thinking, which was still very rough, into words. I also hoped that talking with her about this would help me formulate my ideas better, which it did.

Cathy: One of the most stimulating research projects I have been involved with began in discussions with my husband, an environmental activist and organizer. For years he had suggested that teachers and teacher educators should be doing what environmentalists do when they want to get public attention for their cause: Organize a campaign to educate and inform others to inspire them to activism. As we talked about this notion over a long period of time, it slowly began to take shape in my mind as a potential research project. I started to wonder what teachers could learn from community organizers, and so a few years ago, I seriously started to pick my husband's brain about what the connections might be, thinking aloud with him (usually on long car trips) about how this might turn into some intriguing research. This long incubation eventually began to take shape as a question, and as it did, I started talking to teachers I knew. What were the connections they saw, I asked, and as they started to share with me their experiences, I came to formulate a question more clearly. I then turned, as I often do, to a good friend and colleague in my field to see what she thought about this direction for research. Because she is extremely well read in our discipline, she was able to help me reshape my thinking and articulate my question even further. What I found so vital about this process of working with others at the early stage of my research was both the confirmation from others that my idea was feasible and interesting, and the further ideas that various people, who had very different backgrounds, could provide me.

Our advice to you as new researchers is to be proactive in engaging others in conversation about the ideas that interest you. You may worry that an idea will then be identified as theirs and not yours, but knowledge is by nature social. Both of us believe that interactions play a central role in the creation of knowledge. Because we are all social beings, why not capitalize on that trait as you formulate and fine-tune your ideas for your research, and even as you carry it out?

Writing in a Journal or Log

When we talked about reading, we suggested that you start keeping a log or journal. Journals can also be invaluable tools for helping you externalize and sort through your ideas. As we suggested in chapter 1, if you are starting a research project, consider starting a journal along with it. It is something you can use productively during each stage of the research process. Early on, it can be a place for recording your ideas, impressions, observations, questions, and reactions. Because the early stages of research are about thinking, reflecting, and musing, journals are perfect tools for recording these. Of course, you should also think about your personal habits and how those might influence how you keep a journal. For example, both of us have a habit of thinking of things when we are exercising, showering, or just lying in bed. If our journals are not nearby, sometimes we fail to record our thoughts, and we end up losing them before we can commit them to long-term memory. So even the best intentions can be thwarted by our own limitations. If you are like us, then you might develop a strategy for recording those thoughts that occur at inopportune times or in not-so-optimal settings. We have both found PDAs to be handy tools for this because they are so portable. At times, we have also resorted to keeping a pad of paper close to the areas where we might think of things we want to remember (e.g., Ann has a pad of paper and pencil next to her rowing machine and Cathy has one next to her bed). You can always transfer your temporary notes to your more permanent journal later.

Looking for Recurrent Themes

If you record your responses to what you read, see, and hear, then you'll be well on your way to using this next tool: looking for recurrent themes. What seems to be recurring in what you are reading and/or writing? What ideas keep coming up that interest you? You have to attend to what you are reading and writing in order to pick up on the recurring themes. It's easy to take them for granted, especially if they are somewhat ordinary; in fact, the ordinary themes are the ones that often end up being the most interesting.

You can record the themes that you are noticing in your journal, or you might develop a visual depiction of them. Use a drawing program on your computer, or even

> ## Sidebar 2: *Keeping a Research Journal*
>
> Research journals can take many forms, depending on your own work styles and preferences. Some people, for example, prefer a paper journal (a notebook, binder, or bound book with blank pages). Others prefer an electronic journal, possibly even a blog or other online forum that others can access. There are more options than ever for keeping a journal, and you will want to choose an approach that works best for you (e.g., although you may want parts of your journal to be public so that your advisor, peers, or even research participants can see it, you may want or need other parts to be private). You will also want to give thought to how you organize your journal. Probably you will want different sections for different purposes (e.g., one section for jotting down ideas and questions, one for brainstorming, one for responses to what you are reading, another for frustrations, another for puzzling through interpretations of your data, etc.). Whatever organizational scheme you choose, make sure it's functional—that the information is accessible and easy to find and that it is easy to add to and use.

pen and paper (maybe oversized paper), and visually array the themes you are seeing (you might create a map, a web, a tree, or some sort of graph). You might also record some of the context surrounding the themes; for example, the titles of articles they appear in, the authors who address them, and how they address them. You might think about, and also array, your own ideas in relation to the themes you are seeing. Are your ideas in agreement with the themes you are noticing, or are they contrary to them? Also, what is significant or unusual about the themes, and why are they even themes?

Looking for themes is an important activity throughout all stages of the research process, but it can be especially helpful when you are trying to determine what's important and interesting in your field. Themes will emerge from the reading you do, but they may also emerge from your own observations and early research. In fact, preliminary research is another good way to explore and develop your interests.

Gathering Preliminary Data

Although it may seem premature to do at this stage in your research, you might also start gathering some data. Such preliminary research is different from, but functions a lot like, reading: It deepens your understanding of a topic so that you can think about it in more sophisticated ways and ask useful questions about it. We recommend this tool if you have a sense of the kind of data you might collect and of how you can collect it. For example, you may decide to conduct interviews with individuals knowledgeable in an area that interests you. You might also interview potential participants in your research. Another useful instrument at this stage is an informal survey. A brief, informal survey can be a useful tool for obtaining a quick overview of a situation. For example, if you are a teacher and you are interested in how the other teachers in your school, especially those in other disciplines, use writing in their classes, then you might administer a short survey.

In addition to carrying out interviews or conducting informal surveys, you could analyze some of the discourse produced in a field. As an example, while she was still a graduate student, Ann compared the journal *Physical Review*, which publishes full-length research articles, with *Physical Review Letters*, which publishes much shorter articles designed to disseminate new ideas quickly. Before articulating a research question for her project, Ann looked at both journals. She also obtained pairs of articles written by the same authors and published in both places. She did a quick comparison of these articles to see how they differed. From that, she developed her research question, which sought first to identify and contrast the rhetorical strategies authors used for the two forums.

If you decide to gather preliminary data, we recommend that you familiarize yourself with the various means available for collecting data. We discuss interviewing, surveying, and discourse analysis in depth in chapters 4 and 5. Any of these methods can provide useful insights into your topic. And as you gain experience as a researcher—and greater familiarity with the various research tools—you may find yourself relying on this approach in all of your research.

Broadening and Narrowing

The final tool we wish to address is really more of a strategy that you may find helpful as you define your question. A common problem researchers have when they define a question has to do with scope, so it might be productive to exaggerate your question in both directions—making it too narrow and then too broad—as a way of finding a middle ground. It is also a way to view your question from different perspectives.

New researchers tend to define questions that are too large, so exaggerating in that direction is often easy. These researchers worry that they won't find enough interesting information, so they keep expanding the study or proliferating questions. The latter is something we both have seen frequently: Our students define not one, but several questions, each of which could lead to its own study.

If you do the latter, then you will eventually want to narrow your focus, which need not mean letting go completely of the other questions. You can always write them in your research journal and return to them at a later point in time. You may also want to generate numerous questions deliberately as a strategy for defining the best question for your project. In other words, it can be productive to be prolific with your questions at this point because doing so may help you think of those that could be valuable to your inquiry. What is essential is that you eventually narrow them down: You need to determine what you really want to get at and learn in your research.

Prompt 6: *Broadening and Narrowing.*

Brainstorm a list of questions relating to your topic. Go as broad and large as you care to with your questions. Don't worry about narrowing them or even coming up with a single question. Put the questions aside for a few days. When you return to them, work on narrowing their scope and/or coming up with a single question. You might find it helpful to respond to the following: "By the end of my research, the question I most want to be able to answer is this ... " Try to formulate a single, clear question that captures the essence of what you most want to learn in your research.

Prompt 7: *Trying Out the Tools for Formulating a Question.*

We have now presented and asked you to practice using several tools that can help with discovering and formulating a research question. Select any one of these tools and try using it. Share your outcomes with your classmates. For example, gather some preliminary data by developing a set of questions and carrying out a 30-minute interview with a potential participant for your research. Then transcribe the interview. Or, read four articles addressing a topic that interests you and take notes on the articles. Review your notes and look for recurrent themes. Record and report on these.

Phrasing the Question

Much has been written about the best ways to phrase research questions. Again, because a well-formulated question can influence the way your research proceeds, we believe that it is important at this point to begin thinking carefully about the actual wording of your question. Taking the time right now to compose a well-thought-out starting question can help you establish the focus you want and need for your research before you move on to the next stages of it.

Marion MacLean and Marian Mohr, in *Teacher Researchers at Work* (1999), have one of the best discussions we know of about productive ways to formulate and phrase research questions. They start by suggesting that you take your general topic or problem and write it as three questions beginning with each of the following phrases:

- What happens when … ?
- How … ?
- What is … ?

Each beginning, they say, implies a slightly different approach to the problem. *"What happens when … "* and *"How … "* questions imply observational and descriptive approaches to your inquiry (i.e., "What happens when students are given time in-class to work together in peer critique groups?" or "How do technical writers take their audiences into account when creating documents?") *"What is … "* questions, on the other hand, imply more of a re-examination of concepts in the field, often in more theoretical ways (i.e., "What are peer critique groups?" or "What is usability?").

We have found that asking researchers to start with these phrases can help reduce some of the common problems that occur with the language of questions (e.g., wording questions so the answer is "yes" or "no," or using wording that already implies the answer to the question). MacLean and Mohr (1999) suggest a few additional precautions for formulating questions. First, if your question implies a value judgment, try restating it. They give the following as an example: "'How can I get my students to listen to each other better?' might become 'What is "listening" according to the students in my classroom?'" (p. 6). Second (and this is a problem our students often have), if your question implies the need for a control group, you might consider rephrasing it. For example: "'Does giving students choice about their reading increase their motivation to read?' might become 'What happens when I give my students choices about their reading?'" (p. 6).

Prompt 8: *Phrasing Your Question.*

Working from the drafts of questions you have written in response to earlier prompts, try rewriting your question using the three beginning phrases mentioned in this section: "What happens when … ;" "How … ;" and "What is … ." Do any of these introductory phrases help clarify your question?

LENSES

Bias

It is at this stage of your research that the issues connected to lenses—bias and theory—initially come into play. In fact, we have indirectly addressed concerns related to these issues several times already in this chapter. For example, we have prompted you to think about what's at stake for you in investigating a particular topic and why you are personally invested in it. We have also discussed how being personally invested in your topic is normal, and even good. So, in short, it's natural and quite acceptable for your personal experiences and perspectives to influence your research questions and studies. Consciously considering these experiences, as we ask you to do in the prompt for this section, can help you begin to explore further how your personal experiences and viewpoints might be influencing your research interests.

Prompt 9: *Identifying Bias.*

Reflect on and respond to the following questions:

- What in your own *personal experience* has influenced your thinking about the issue you are considering for your research? In what ways do you think your experiences are influencing your thinking about the issue?
- How are your *personal beliefs and preferences* affecting what you are thinking of doing in your research? What are those beliefs and preferences?
- What do you think might be some *biases* you have in your approach to your topic?

When you have a personal stake in your research topic, it is inevitable that some amount of bias will influence what you do. However, bias is not always bad, especially if you make yourself aware of it as you carry out your research. In other words, everyone has biases and, frankly, they are not something that can be eliminated easily. In fact, composition scholars have increasingly addressed the issue of bias and its place in research (e.g., Borland, 1991; Kirsch, 1997; Kirsch & Ritchie, 1995; Moss, 1992; Yagelski, 2001). Many of these scholars talk about bias in terms of personal subjectivities that arise from our own perspectives and that influence how we view and interpret situations. They argue that subjectivities are always present in our research. What we need to do, they say, is reflect on and acknowledge them. Such reflection and self-awareness, both in your research and in your writing, can help you recognize when these subjectivities might be interfering with your research. No research is perfectly objective, and it doesn't need to be. Your personal perspectives will always color what you see, how you see it, and how you interpret or make meaning of what you see. There is no way around that, and that's why you will never carry out a research project in exactly the same way someone else does, or come up with exactly the same findings, even if you both started out with the same question. Because of the importance of personal subjectivities in research, we continue to address this issue throughout the text.

Theory

The other aspect of what we are categorizing under lenses is theory. Theory is sometimes portrayed as nebulous and esoteric, something that's beyond new researchers and only appropriate for certain prominent scholars. We do not see theory in that way; in fact, we see it as probably your most important ally in research. It supports what you do, and it is what allows you to make meaning of your research. It is the scaffolding for your inquiry, and it influences the question you formulate, the approach you take in carrying out your research, the way you interpret your findings, and even the way you present your findings.

Prompt 10: **Reflecting on Theory.**

If you are reading this text as part of a research class, then you should take a moment to reflect on what theory means to you and then share your perceptions with your classmates. You can use the following questions as prompts:

- What is theory?
- How does it influence you and what you do? (You might think about the latter in both personal and professional terms.)
- What are some theories in your field(s) that have influenced you?

For us, theory is a way of viewing something, a perspective on or an explanation for some behavior or activity. Consider some examples. In the field of writing, broadly defined here, there are numerous theories that may influence your scholarship—both in terms of how you understand writing and how you understand the process of research. For example, Vygotskian activity theory and its derivative, situated learning theory, offer an explanation for the ways in which learning occurs. These theories suggest that humans learn by doing, by engaging in some kind of authentic activity within a domain. Individuals start out at the periphery of a domain and gradually, as they gain more experience and familiarity with its activities, move toward its center (this process is referred to by Lave and Wenger, 1991, as legitimate peripheral participation). Social construction theory, which underlies the view of research we present in this text, suggests that knowledge is constructed through social means, that it does not simply exist out there as something to be discovered, but is something discussed, negotiated, and debated. This theory is especially popular in writing research because most writing scholars view knowledge as rhetorical and see discourse functioning centrally in the generation of knowledge.

Other influential theories in the field of writing include writing process theories, which suggest that writing is a recursive process and a complex social and cognitive activity that can be taught and learned; romantic theories, which view writing as fairly esoteric and personal (according to these theories, great writers are born or inspired); and current traditional theories, which view writing in more mechanistic terms. Classical and neoclassical or new rhetorical theories offer additional ways of viewing discourse, focusing on the contexts for and the situations surrounding discourse. Postmodern theories help explain the personal and contingent nature of knowledge and how personal subjectivities influence all knowledge. These are just a few of the common theories in our field. At this point, you might use the prompt for this section as an aid in identifying some other influential theories.

Prompt 11: *Identifying Theories.*

Read three articles related to your topic and try to identify the theories that underlie the authors' claims. Also, after you read all of the pieces, reflect on them and on what you learn about theory from each of them. What role does theory play in each piece? How does it function in each piece?

What is important here is not that you construct an exhaustive list of all the theories that influence writing research, but that you consider what theory is and what it means to you, and that you begin reflecting on how various theories may influence your research. In our view, everything is theoretical because everything has some explanation that can help you understand it better. Theory exists by virtue of the fact that there are various ways of viewing the world, and those ways often are highly individual. In fact, just as your personal interests may influence what you decide to research, so might your personal theories (see the final prompt for this section). Theory is something you will want to think about at every stage of the research process. It also is something that will influence and help you with each stage. The sooner you begin thinking about it, therefore, the more it can help you.

Prompt 12: *Identifying Personal Theories.*

Identify some personal theories or ways of viewing the world that you think are influencing your topic for your research. For example, maybe you believe that you can learn more from looking at the process by which individuals compose texts than by looking just at their products. As a result, you may decide to observe your participants as they compose instead of simply analyzing their written products. A variation of this prompt is to consider how your personal view of the world or, on a smaller scale, of your field or subject area, contributes to your theoretical perspective. If you believe, for example, that texts are influenced by a number of social inputs, how does that affect your larger theoretical perspective?

ETHICAL ISSUES IN FORMULATING RESEARCH QUESTIONS

Other issues about which you should begin thinking, even this early in the research process, are ethical issues. For us, research ethics relate most directly to how our research affects our participants and the settings we study, that is, the way we formulate a question may imply certain beliefs about the participants in our research setting and may, in fact, make unintended judgments about our participants. In a study that Cathy conducted when she was in graduate school, for example, she began with a question about how "process writing" was taught across the school day for one 10th-grade student. Even asking that question implied a certain bias (that process writing should or would be taught) and implied certain judgments about the teachers in the school (who was and who was not approaching writing in that way). As a researcher who would "live" many months in that school setting, Cathy had to consider carefully the ethics of that research question: how it might affect the work of the teachers and how they might respond if they perceived that the question carried certain judgments about their work.

Ethical issues also come into play with individuals and settings that are influenced by our work even if they are not directly involved in it. In Cathy's case, these individuals included the students in the classes she would eventually observe, the principal of the school, and even upper level administrators. There is a disposition associated with being ethical, and both of us believe that it is easier to acquire and maintain that disposition if you begin considering these issues early on. Too often ethics is something we consider after-the-fact in research rather than making it an integral aspect of our research, even as we are just formulating a research question. As an example, answering your question may entail altering the normal activity in a setting. Your question may also have implications for your research participants (e.g., putting you in a position of power over them). As much as possible, you should begin considering the ethical dimensions of your research as early in the process as you can. Some aspects of research, like human subjects review (see chap. 3), will prompt you to think about ethics explicitly. However, a lot will be left to you; therefore, the more you can get into an ethical mind-set, and the earlier you can get into such a mind-set, the more prepared you will be for just about any ethical issue you might encounter.

Prompt 13: *Identifying Ethical Concerns.*

Consider your research question and reflect on the ethical issues it suggests. What kinds of ethical concerns might be raised in the research suggested by your question?

The strategies you use to formulate a research question may also raise ethical issues. For example, if you carry out preliminary observations, you will need to think about how you will use your observations and whether you need to inform the people you observe. If you are just trying to get a feel for a situation, it may be fine to simply observe without informing anyone. However, if you plan to report your observations, and if they implicate or identify anyone, then you definitely should seek permissions. Online observations especially raise ethical issues. Although it may be appropriate to lurk in an online discussion if you just plan to generalize from it, you should seek permissions if you intend to use what you observe or read (Gurak & Silker, 1997). In any of these situations, online or in person, you need to think about how you will use the information you obtain and how your participants may be affected by your research. Unfortunately, these things seldom are straightforward; therefore, in every situation, you need to make personal judgments about what is fair, responsible, and honest.

Another aspect of being ethical is acknowledging your biases and thinking about how those might influence your research. Here you should be honest both with yourself and with others. Identify and articulate what perspectives are influencing you and how they are influencing you. Again, bias isn't always bad, but if it goes unacknowledged it can limit how you think about and approach your research, which may have ethical implications.

At this point in the research process, we offer the following suggestion: Be aware both of your own ethical stance and of the inherent challenges to that stance that your research might pose. Take time to write about your concerns in your research journal and consult with other experienced researchers about those concerns. Ethical concerns need to be embraced and acknowledged both at the outset of your research and through each stage of it. For this reason, we will say more about ethics, and about the ethical considerations associated with the various stages of the research process, in all of the subsequent chapters.

CONCLUSIONS

You probably didn't realize that there was so much to do at such an early stage of the research process. However, if you articulate a good question, you will be well on your way to a successful project. And because research is a process, you will very likely adapt and fine-tune your question as you begin learning, encountering, and discovering new information through your research. (We have included in the appendix of this chapter some examples of research questions that two of our students have articulated. We show successive drafts of those questions so you can see how they changed. We also share their reflections on the changes.) We encourage you, as you begin formulating your question, to keep every draft of it. Record each version in your research journal or keep an electronic or paper file of your drafts.

We have one final suggestion to offer before you move to the next chapter and to the next stage of your research. Now is as good a time as any to begin thinking about who might be interested in your research and for what reasons, and who your research might influence and affect. If you are carrying out your research for primarily personal reasons, then this may not be as important. However, you may decide to share your findings with others so that they too can benefit from what you have learned. If you are a teacher, for example, a primary concern you might have is with sharing your findings with your colleagues and learning from their experiences as well. So, here are some questions to consider as you begin your journey through the various stages of the research process:

- Who will care about the research you do?
- Who will your research affect?
- How will your research affect your own situation?
- What other situations or contexts will your research affect?
- Who does your research need to matter to and why?

Every researcher will respond to these questions differently. In fact, you may respond differently during each stage of your own research; it will depend on where your research takes you.

APPENDIX: SAMPLE RESEARCH QUESTIONS

Debra Fitzgerald, July 2004

Research Question Draft 1: What would a high school curriculum that successfully integrates literature study and multigenre writing into a single course look like?

Research Question Draft 2: Assuming multiple genre writing improves student writing, what would a high school curriculum that successfully integrates literature study and writing in multiple genres into a single course look like?

Research Question Final Draft: What kinds of writing assignments would successfully integrate interpretive literature study and writing in multiple genres, with the end product ideally demonstrating both high level interpretive skill and writing skill?

Narrative/Commentary: As I began to develop a research question, I knew I wanted to look at writing in multiple genres—almost as a sort of protest against the often-immovable force of the traditional five-paragraph essay. But I'm a literature teacher, so I wanted to integrate literary analysis into my study of student writing. My first research question dealt with both of these issues, but the phrase "high school curriculum" seemed too ambitious. I really didn't want to write a curriculum; I wanted to look at real student writing and see what kinds of assignments help them think about literary works in sophisticated ways. My second research question had that pesky little "assumption" in it, which I felt wasn't appropriate because I was assuming as fact the very premise I hoped to demonstrate with my research! I knew I wanted my question to address the companion issue of improving student writing, but without that assumption up front. Ultimately, I had to ask myself what it was that I really wanted to know at the end of my project. It sounds obvious enough, but it's easy to lose sight of that simple precision when developing a study. I decided that I wanted to learn what kinds of writing best help students interpret literature in increasingly sophisticated ways, while also helping them develop sophisticated writing skills. I think my final research question asks that, but if I thought about it some more I'm sure I could make it even more precise.

Erica Pilon, July 2004

Research Question Draft 1: Is there an expected or preferred interface for electronic communication and what can technical communicators learn from their users' expectations? How do user biases and document design standards affect these preferences?

Research Question Draft 2: Is there an expected or preferred interface for electronic communication and what can technical communicators learn from their users' expectations? How do user biases and document design standards affect these preferences? (Same as draft 1.)

Research Question Final Draft: Do end users prefer Windows or web interfaces for practical applications? How do the ways that users learn and their prior experiences seem to influence their preferences? What are the implications of this for the work performed by technical communicators, especially in the area of interface design?

Narrative/Commentary: A problem I am having is that my research question does not cover all of the literature and theories that I discuss in my proposal .... I am not sure why I was so married to the original wording of my research question. I had some great feedback on my research question at the beginning of the semester, so I think I wanted to build on that. But, as I delved into the reading and literature around my topic, my interests started to move in a different direction. I just [was not] able to articulate that explicitly in a revised research question, primarily because I was stubbornly trying to build my research on the original question. I am kicking myself right now because I feel like I could have been more productive if I had been more flexible with the question earlier on. But, that is one of the great suggestions that evolved from our workshop. ... And I think a little light bulb turned on in my head as a result. I have subconsciously been revising my outcome all semester. Now that I am aware of the problem, I can rethink my research question and try to streamline this project into something manageable.

How Do I Find Answers? Planning Your Qualitative Research Study

By this point, hopefully you have developed a question to focus your study, which means you are ready to think about the next stage of the research process: finding answers to your questions. This chapter offers strategies that will help you plan a qualitative research study. We address how to write a problem statement and expand that into a research proposal; how to review the literature; and how to get started with your research, including how to select research sites and how to obtain access, permissions, and human subjects approval for those sites. In the two chapters that follow, we extend this discussion to carrying out your research (e.g., how to observe a setting, how to prepare for and carry out interviews, how to design and administer a survey or questionnaire). In short, we present in this and the next two chapters the nuts and bolts of planning and carrying out a qualitative study, the tools of the trade that are essential for finding answers to your research questions.

GENERAL ISSUES FOR PLANNING A RESEARCH STUDY

The following general issues are discussed in this section:

- Planning a study that will answer your question
- Planning a study that isn't too large or too small

Planning a Study That Will Answer Your Question

Planning is one of the most important activities in the research process. It allows you to prepare and get yourself ready to carry out your research. If you have a good sense of your research question and are enthusiastic about it, you may be tempted to dive right in. However, it is important to have at least some sense of *what* you intend to do, *how* you will do it, and *where* and *with whom* you will do it. That is not to say that your plans won't change, but a good plan will focus you and also help you if you do change something. The simplest analogy here is a road map—you want to begin your research journey with some sense of where you are heading and how you will get there. At the same time, you also want to be able to choose alternative paths.

Although the tools offered in this chapter will assist you in your planning, your research question is the most obvious starting place. Most decisions about how to proceed with your inquiry will be based on it. Your question is also important for thinking about the scope of your research. Our advice is to fine-tune your research question as much as you're able to and then use it as the starting point for thinking about all of the issues you need to consider in undertaking your project. Then, as you work your way through this chapter, keep adding the new issues and considerations to your plan. Eventually, the plan will become your full-fledged research proposal.

Planning a Study That Isn't Too Large or Too Small

One particular issue to consider as you begin planning your research project is its size. Often, especially if you are just starting out, the tendency is to think big. You may do this because you are afraid you won't get everything you need or that you won't have enough to say. Isn't it better, after all, to research extensively so you don't run out of things to write or so you don't overlook something that seems obvious to others? However, if you try to do too much, then your research can get away from you and/or end up taking much longer than necessary. For example, if you are doing a master's thesis, you are probably earning the equivalent of three academic credits for it. That means, technically, that it should be something you can complete in one semester.

But scope is difficult to define, even for experienced researchers. What is too large or too small? Your research question has a lot to do with determining that: Your primary concern should be with investigating your question thoroughly without biting off more than you can chew. The time frame you are working with should also be a contributing factor. The best advice we can offer here is to talk to more experienced researchers—get feedback from your project director, for example, on whether what you're planning seems to be too much. Also, be realistic and keep a sense of perspective. For example, we tell our students to make their mantra, "It's just three credits." Keep in mind too that you can always adjust the scope of your project as you go. The recursiveness of the research process facilitates this; however, determining and adjusting the scope of projects takes practice, so it is important to begin thinking about scope as early as possible. And remember that as you develop your skills as a researcher, making judgments about scope should become easier.

Some of the tools and strategies discussed in this chapter will prompt considerations of scope. As you use these tools, think about scope by asking yourself the following: How doable is what I'm proposing given my time frame? Is what I'm planning to do enough? How realistic is it? Could I do it any other way? As you consider these questions, think about both your research question and your goals.

Prompt 1: *Drawing a Road Map.*

One strategy you might find helpful as you plan your research is to draw an actual road map for your study. Find some oversized paper and some markers or colored pencils. Place your question in a prominent spot and then let your creativity take over and draw what you plan to do for your study. If you are not quite sure of your plan, begin drawing your map now, but fill it in as your plans develop. Include in your drawing every aspect of your research plan that you can think of, including personal aspects like rituals, breaks, obstacles, and the rewards you will give yourself as you achieve certain milestones.

TOOLS AND STRATEGIES FOR PLANNING A RESEARCH STUDY

The following tools and strategies are useful when planning a research study:

- Writing a research proposal
- Writing a problem statement
- Reviewing the literature
- Selecting a setting for your research
- Obtaining permissions to study the setting
- Obtaining human subjects approval

Writing a Research Proposal

Your research question provides a starting point for planning your research, but your proposal elaborates your question by presenting your plan for investigating it.

It is also the culminating piece of your planning. However, we talk about the research proposal first in this section to give you a sense and a vision of where you are headed. Keep in mind that all of the tools and strategies presented in this chapter will help you to complete the proposal.

In the proposal, you select and lay out the tools and strategies you intend to use in your research. This stage, then, entails forging connections between your research question, its larger context (which we say more about shortly), and your plans for investigating it. In fact, researchers at all levels of expertise typically write some form of a research proposal. These documents contain specific types of information, and thinking through this information can provide a productive starting point for planning your research.

Research proposals[1] usually contain the following:

- A clear statement of the research question (this is sometimes presented more elaborately in the form of a problem statement).
- A literature review that places the research question in a larger context and establishes the gap in the literature that the question addresses (your problem statement, addressed in detail in the next section, provides a foundation for the literature review).
- A statement of the researcher's personal interest and stake in the project.
- A statement of the goals and objectives for the research.
- A description of the researcher's plans for carrying out the research (essentially the methods the researcher will use; you should take care to connect these to your research question and to provide sufficient information about why and how the methods you have identified will be used).
- A description of the audience for the research (to whom it will be targeted and on whom it will likely have an impact).
- A statement of the anticipated outcomes for the research (e.g., what you will produce—research article, thesis, dissertation).
- A statement indicating whether approval for using human subjects has been obtained, or when it will be sought.

Research proposals, in general, should be matter-of-fact and realistic. They should also be focused, coherent, organized, sufficiently detailed, and complete. Proposals, most of all, should be persuasive. You want to convince your audience—your thesis or dissertation committee, funding source, even yourself—that you have a meaningful research question and a clear idea of how to investigate it.

---

[1]Research proposals can function in several ways (e.g., they may fulfill an academic requirement or they may be written to obtain funding). The content and organization of a research proposal is typically specified by who is requesting and receiving it. In some cases, the information that's needed is specified in a formal document called a request for proposals (RFP). Researchers need to follow RFPs precisely or their proposals may not even be considered. Academic programs may also have guidelines for writing proposals. You should familiarize yourself with such requirements before you start your proposal.

The sections that follow discuss the tasks you should undertake as you prepare to write your research proposal. Remember that the research process, including the planning aspects of it, are recursive, so there is no absolute order to these tasks. The order we present is simply intended to be suggestive.

Prompt 2: *Developing a Criteria Sheet for Your Proposal.*

Use the list of information typically included in a proposal, and the characteristics we have defined for them, to develop a checklist for evaluating your proposal. Continue adding characteristics as you read through the strategies in this section and as you consider the requirements of your program and your own needs. If you are developing a proposal as part of a research class, you might also develop a common checklist.

Prompt 3: *Planning Your Research.*

Alternative 1

If you had just one semester to complete your research, what would you do? How would you set up your project and carry it out?

Alternative 2

Swap research questions with a partner. What seems, from your perspective, to be the best way to find answers to the question(s) your partner has posed?

Alternative 3

Begin developing an idea map or tree diagram for your research proposal. Use your research question and the bulleted list of contents we present to help you generate your ideas. You probably won't have all of these details worked out yet, but you may be surprised at how much you do have.

Writing a Problem Statement

Once you have settled on a research question, you'll want to begin situating your question in a larger context. That larger context includes both the reading you have done that addresses your research question and the more personal context of your

Sidebar 1: *Selecting a Thesis Committee/Director*

If you are a graduate student, you need to decide—if you haven't already—whom you will ask to direct your research, and/or whom you will ask to be on your committee. In making these decisions, consider the following:

- Which faculty members work in the area you plan to research, and/or what can various faculty members contribute to your work?
- Which faculty members have work styles and personalities that are compatible with yours?
- Which faculty members can challenge you in the ways you want or need to be challenged?
- Which faculty members can give you the amount of guidance and support you feel you will need?

Make appointments and talk to different faculty members about your research: Let them know what you plan to do, how you plan to do it, and what your anticipated time frame is. Find out if they are interested in your project and in working with you. Also find out how available they are (find out, e.g., if they are available when school isn't in session—thesis and dissertation projects seldom conform to the academic calendar). If you need to select a faculty member outside of your department, which many programs require, select individuals whom you believe will make meaningful contributions to your work. In addition to telling them about your research, inform them of the procedures your department follows for thesis or dissertation projects. You might also invite your director to join the conversation, especially if the person has not previously served on a committee in your program.

own interests and motivations for articulating the question. A good first step for embedding your question in these larger contexts is to create a problem statement. Problem statements connect your research question to a larger context and show the gap your question addresses. They establish the uniqueness of what you are doing and the contribution your research will make. (See the sample problem statements included in appendix A).

In other words, your problem statement does the following:

- It establishes how and why the issue you are investigating is a problem, both for you and for others in your field.
- It addresses the importance or significance of the issue, again, both for you and for others in your field.
- It provides background that addresses the field's investment in the issue as well as your own investment in it.
- It addresses what has already been done and what others have already said about the issue, as well as what has not been done or said about the issue.

So, while your research question guides your inquiry, your problem statement, which usually can be written in a page or two, establishes the significance of your question and connects it to other research in your field, as well as to your own interests. It provides a backdrop for the plan you will develop for your research, which you will elaborate on in your research proposal.

Prompt 4: *Writing a Problem Statement.*

This prompt has two parts. First, read the problem statements in appendix A and discuss whether and how the four characteristics in the bulleted list are presented. Second, expand your own research question into a problem statement using the bulleted list. Share your problem statement with your classmates and use the feedback you receive to determine whether and where you need to elaborate it.

Reviewing the Literature

If you are developing a study as you are reading through this text, then no doubt you have already begun reviewing the literature. This activity takes on renewed importance as you plan your research, and the outcome of your efforts typically will be presented in your problem statement and proposal and, eventually, in the final write-ups of your project. Reviewing the literature is an ongoing activity in research, so much so that we had difficulty determining where to discuss it in this text. We chose to discuss it at some length here because this is the stage at which you're most likely to begin undertaking this task in a systematic manner. Now that you have a fairly good idea of what you plan to research, you want to find out who has addressed it already and what they have or have not said about it. You want to begin situating your work in the larger field, relating it to and positioning it in the conversation of that field.

If you are fairly new to this activity, then you might first consider the questions in the prompt for this section. Our own students often admit being anxious about this part of the research process. They express concerns about not being able to find anything, finding too much, finding the one right thing, not knowing where to look, and not knowing how to make good judgments about what they do find. Such concerns, and others you might have, can make this part of the research process seem challenging. Typically, there is a lot of information out there, so you can never really be sure if you have found everything that is relevant. However, you can do several things, which we address in this section, to ensure that you carry out your literature review in a productive and effective manner.

Prompt 5: *Completing a Lit Review.*

Consider these questions:

- What concerns do you have about doing a literature review? (Try to list everything you can think of, including concerns about consulting the right databases, about finding things that are relevant, about not finding enough, and about technology.)
- How do you view yourself and your own work in relation to the scholars whose work you are likely to encounter in your literature review?

The second question in the prompt for this section has to do with developing your identity as a researcher. Part of the process of becoming a writing researcher is learning to make a contribution to the conversation in your field, which many students find challenging. Both of us, for example, have heard students express doubts about contributing something meaningful to their fields. The students we interviewed as we were writing this text indicated that they saw the scholars they were reading as being beyond their reach, as experts with special credentials. In some respects, this is true. However, your ability to contribute to the conversation in your professional setting or to the larger scholarly conversation in your field, if that's your goal, may be greater than you think. Again, we hope the strategies presented in this section will help you realize this. As a first step, you might find it helpful to complete the next prompt, which will begin familiarizing you with the professional journals in your field: what they are, what kinds of articles they publish, and what the requirements are for publishing in them.

Prompt 6: *Reviewing a Journal.*

Select a professional or academic journal in your field and review three recent issues. Produce a write-up that contains the following information:

- The title of the journal, the name(s) of its editor(s), and the specific issues you reviewed.
- A general description of the topics addressed in the journal.
- An indication of the genres published in the journal (e.g., essays, research reports, case studies, narratives, etc.) and their average length.
- An indication of the types of research published in the journal (e.g., ethnographic, experimental, case study, survey, etc.).
- A short summary (two or three paragraphs) of one article from any one of the issues you reviewed.

Share your write-up with others in your class. What similarities do you notice among the journals? What differences?

Where to Begin

If you are not familiar with electronic databases, or if it has been awhile since you've done any sort of library research, then a good place to begin is with a refresher course or session with a librarian. At the very least, familiarize yourself with the electronic and other resources available at your library (e.g., their journal holdings). You can access most libraries' electronic resources by logging onto your campus' Web site and then clicking through to the library's site. If you are not familiar with that site, take some time to explore it so you can see what's there. If you have a student account, you can also usually access the databases to which your library subscribes. In other words, you can do this directly from home, or from wherever you might be accessing the Internet, rather than having to physically visit the library.

An electronic database is essentially a searchable collection of resources. There are literally hundreds of databases, and your library will likely only subscribe to a selection of them. Most of the major ones you might consult will be available, and for those that are not, you can check other libraries. Basically, your school's library pays a fee to the company that maintains the database and, as a result, obtains access to that database for its patrons. All databases are limited in some way. Some focus on particular subject areas (e.g., Education Abstracts, ERIC, Gender Watch, Health and Wellness Resource Center, JSTOR, Lexis-Nexis Academic Universe, MLA Bibliography). Some focus on types of publications or particular genres (e.g., Dissertation Abstracts, Essay and General Literature Index, HarpWeek, *New York Times*, Poem Finder, Short Story Index). Some focus on coverage (e.g., Electronic Collections Online, netLibrary, NewsBank-NewsFile, Newspaper Abstracts, Project Muse, Readers' Guide Abstracts). And some focus on reference and other types of information (e.g., Facts.com, FactSearch, *Oxford English Dictionary*).

Prompt 7: **Completing a Lit Review.**

Identify three online databases that might be useful to your literature search and review them. Make note of the following:

- Their purposes.
- The interface they use (including fields and prompts).
- The search strategies they recommend or lend themselves to.
- What you are able to find with them.
- What recommendations or tips you have for using them based on your experience trying them out.

For those of us who remember card catalogs and paper indexes, which now have vanished from most libraries, the electronic resources are a welcome alternative. (Although some individuals admit missing card catalogs and the large indexes, most of us have adapted to the new technologies.) These resources are usually very easy to access and search, although it helps to have effective search strategies, which we address later. Also, whereas some databases provide just citations and some offer abstracts, many databases now also offer the full text of the articles themselves. The full text is a great bonus—you can download and save and/or print the article—but it also has some limitations. If the full-text file is formatted in hypertext markup language (HTML), which many are, then the original formatting of the article is lost. This means that if you quote or cite the article and need to note page numbers, you cannot do so accurately without retrieving the original publication. If the article is a portable document format (PDF) file, on the other hand, the original formatting will be preserved and you can cite it accurately. You need to be aware of the differences between these and know which of them you are downloading.

How and Where to Look and What to Look for

When you begin searching online, the first thing you need is a list of key words. These probably are the most critical aspect of your search strategy, so you should take your time brainstorming as many as possible. Start with the key words in your research question and then think of others that might apply. For example, if you are investigating how computers affect revision in middle school classrooms, then you would probably begin with the words *revision* and *computers* (or *technology*) and go from there. Once you have generated a list, you should then go through and select those that seem especially relevant (e.g., with the computers and revision study, you might finally decide on some of the following: *revision, computers, technology, computers and writing, middle school English*). If your list of key words becomes too long, or if all of them are too broad, then you will need to limit and focus your list. Otherwise, your search will uncover too many sources that are not relevant to your topic. For example, if you simply type "computers" into the search field in ERIC, you will end up with more than 21,000 entries, most of which won't be relevant. Most library Web sites have tips and instructions for generating key words that you might find helpful. See, for an example, http://www.lib.duke.edu/reference/catguide/keyword.htm

Although you need a well-chosen list of key words to carry out an effective search, you also need strategies to limit your search. One way of limiting your search is to use Boolean search strategies—if a database allows you to. Boolean searching is based on Boolean logic. It is named after the British mathematician George Boole and entails using logical operators ("and," "or," "not") to narrow your search. The operators establish the relationships between items. Some databases prompt you for them, and some fill them in for you. Others require that you fill them in. (As with key words, most libraries have online tutorials or instructions for searching that describe how Boolean logic and operators work.)

If you are researching writing, you will likely find that some databases are more helpful than others. Some of the more common ones used by scholars who do writing research include Education Abstracts, ERIC, MLA Bibliography, and Firstsearch. Other databases you might find useful include ArticleFirst, Arts and Humanities Search, Dissertation Abstracts, GenderWatch, Humanities Abstracts, Internet and Personal Computing Abstracts, Kraus Curriculum Development Library Online, Project Muse, and Social Sciences Abstracts.[2]

To search for books, you need to search your own library's holdings using its online catalog. Your library may also have agreements with other libraries (e.g., with other university libraries in your state) and, as a result, you may have access to their online catalogs as well. If your own library does not have a particular book in its collection, you should be able to obtain it through interlibrary loan. This also applies to paper and microfiche copies of journals, which you may need to obtain if you cannot locate full-text copies online or if you need the original formatting and pagination of an article.

What to Do With What You Find

Once you have discovered a number of sources for your work, you will need to decide what to look at and use. You will likely uncover many more sources than you can possibly read, so how do you narrow those? In addition to having good search strategies, it also helps to have some strategies for evaluating what you find. We recommend that you ask the following questions:

- Is the source credible? Is it a reputable publication (journal or book) published by a reputable organization (e.g., a professional association or major publisher)?
- Is the source recent? (This question really is relative as some older sources may be pertinent to your study. In some research, however, older sources are not useful.)
- Is the source relevant? (This is a trickier question that begs the larger evaluation question. Relevance depends on your research question and on your approach to your research. It ultimately is something you need to determine.)

---

[2]Whereas general search engines can provide some useful leads, they will not yield a comprehensive list of the academic literature on a subject. For that you will need to use the more targeted databases, which are constructed to provide just that sort of information.

After you have evaluated and selected the sources you plan to consult, you then need to spend time reviewing them, which will likely lead you to additional sources. Reviewing the literature is an ongoing process. You need to set practical limits so that you can keep making progress with your project. You also need to determine what you can return to later, knowing that you will continue to look for and read sources throughout your project. Keep in mind too that you may not need to read every source, especially books, from start to finish. You may find that just certain parts of a work are relevant to your project.

How to Compile and Write Your Literature Review

At some point, then, you need to set limits for your literature review so that you can begin writing it up as part of your proposal. Usually, the lit review is that part of your proposal where you situate your work in relation to what else is out there; you build a context for it and show the gap that your own work addresses. You do this in your problem statement too, but the full literature review extends that. If you have collected a large number of sources, then you need to determine how best to present these, both in relation to one another and in relation to your own work. Appendix B contains two research proposals that were written by our own graduate students. These two proposals, we believe, effectively demonstrate how to integrate a literature review into the plan for your research. (We will refer to these proposals again in the next two chapters where we discuss methodology.)

In other words, a literature review should not simply list all of the sources you reviewed (Smith said this, Baker said that, Rouch said that …) without relating what you are reading to what you are doing for your research. This sort of lit review ends up reading like a laundry list. In contrast, you should integrate your lit review by grouping and relating the items you present and connecting those items to the other parts of your proposal. This makes for a more coherent presentation.

To achieve this coherence, ask yourself the following questions:

- How does what you are reading connect to what you are asking and to the other things you are reading? (You should ask this for every piece you read.)
- How is each piece you read relevant and meaningful? How does it fit into the larger conversation about your topic?

You might even write out some of your answers to these questions, or array them visually. You can use these representations then to establish relationships when you write your review.

As alluded to in this section, we always urge our students to see the lit review as an integral part of their research study—and not merely as a hurdle designed to demonstrate to an outside audience that they have done their homework. We believe a lit review is not about finding hundreds of sources, and it is not about summarizing sources not connected to your research question. Rather, it's about understanding the prior conversations that exist about a topic in order to see how your study might extend those conversations. The lit review is a place where you can begin to see how you, as a researcher, have the potential to be a part of the ongoing discussion that defines your field.

Selecting a Setting for Your Research

Another significant step in planning a qualitative research project is selecting the setting for your research, which involves determining how you will gain access and permissions to study that setting and whom you will involve in your study—that is, who your research participants will be.[3] In some situations, these decisions are quite simple—they are determined by the type of research you plan to do, your purposes for the research, or the question you are asking. For example, if you are a teacher interested in examining the planning strategies of high school students, your setting will probably be your own or a colleague's classroom. You will negotiate entry by seeking approvals from your school's administrators, school board, and/or parents. If the purpose of your research is to determine the effectiveness of change communication strategies in the company at which you work, then the setting will be the company, and the approvals will come from the company's management (e.g., from the company's CEO and from managers of the departments that were affected by the change communication). And, if your question seeks to determine how engineering coop students learn to write in school and in the workplace, then your settings will be both their classrooms and their workplaces. Here you would need approvals from their teachers as well as from the students' supervisors in the workplace. In all of these scenarios, because you would be working with human participants, you would also need human subjects approval, which is discussed shortly.

In many situations, decisions about your setting and participants are not obvious. In these cases, you need to think, first, about the setting that would be best for exploring your research question. Once you have identified a potential setting, you then need to determine what might be entailed in obtaining access to the setting. This includes thinking about things such as seeking permissions and agreeing to

---

[3]We use the term *participants* as opposed to *subjects* because the former suggests a much more collaborative relationship between researchers and the individuals with whom they interact during a study. The term *subjects* suggests a hierarchical and controlling relationship in which the authority and power lies with the researchers. The general move in qualitative research, especially with the influence of post-modernism, has been away from such conceptions of researcher and researched toward notions such as that of co-researchers, a notion addressed in chapter 7.

any conditions they might set (e.g., you might not be allowed to review some documents or speak with certain individuals at the site).[4]

In choosing a research site, you should also consider practical and even personal issues. These include how easy it will be to get to the site, how comfortable it will be for you to be there, and how interested the people there will be in working with you. Sometimes convenience and expediency become important factors in choosing a site. For example, one setting may seem more ideal than another given your research question, but a less ideal setting may be more accessible, have fewer restrictions, and simply seem more comfortable for you to be in. You need to determine if you can or should make the trade-off.

All of this is basically to say that many things will influence your considerations of where to carry out your research and with whom. Perhaps Ann's experience, described in the sidebar, will help clarify at least some of what is involved in planning this aspect of your research.

Obtaining Permissions to Study the Setting

Once you have selected a setting, you need to begin negotiating access and securing all of the permissions you need to carry out your research in that setting, even if the setting is your own workplace. First, identify the individuals who have the authority to make these decisions. Find out what they need from you. Sometimes they simply want to meet and talk with you about what you will be doing. Other times they ask for documentation, such as your resume or vita, your research proposal, or a specific research plan that indicates how you plan to use the setting and how you will interact with your participants. You may also need to provide examples of your work or offer references. If you are conducting research in your workplace, they may also want reassurance that the research will not interfere with your own work or with the work of others. All of this is meant essentially to help these individuals (the settings' gatekeepers, in essence) determine how competent and professional you are as a researcher—even if they already know you as an employee.

As an example of the latter, when Ann carried out her research with physicists, she frequently was asked to provide copies of papers she had written. The physicists also often questioned her about her academic background and prior work. They wanted to see what she had done and how well she knew and understood her own and their field. She felt, on many occasions, like she was being tested: She

---

[4]One of our students, because of the highly confidential nature of the projects at her workplace, was told that her project readers needed to sign confidentiality agreements. She too had to sign an agreement indicating that the project would not be read by anyone other than her committee. She still opted to do the project because of the benefit it would have for her in relation to her job functions. Another student, who was using her company's software in her project, was asked to have all of her participants sign release forms promising not to reveal anything about the software to anyone outside of the company.

Sidebar 2: *Choosing a Research Site (Ann)*

As we were writing this chapter, I was involved personally in making the kinds of decisions we discuss in this section. I was about to embark on a new research project. I had defined my questions for the project and even written a proposal for it. The purpose of the latter was to secure a year-long sabbatical leave from my university so that I would have ample time to pursue my project. The one aspect of the project that I hadn't determined yet, however, was the specific settings in which I would carry out my research. I knew generally what those settings should be—a life sciences research lab in a university and a for-profit life sciences company, either a pharmaceutical company or a small organization whose main purpose is to carry out work in some aspect of the life sciences. Part of the reason I hadn't made my final decisions is that I wanted there to be some link between the academic and the professional settings (e.g., students from the academic setting working in or carrying out internships in the professional setting). I also, quite honestly, was procrastinating because of some anxieties I was having. I was worried, for example, that the professional settings would not be receptive to my work and that the proprietary nature of their work would impede my research—for example, I would not be permitted to analyze certain kinds of documents because of concerns with confidentiality. (My research is concerned with genres and with the decisions that attend their adaptation, especially the decisions tied to the specific culture of the organization and to its goals. Therefore, I needed access to the genres used in the organization, as well as to the individuals who were writing and adapting the genres for their specific purposes.) I also was worried about choosing a good, or "the best" setting—one that would yield lots of interesting findings and data as opposed to one that might not be as interesting or "sexy." I also had the unfortunate experience of seeing two of my potential sites go out of business when I was in the early stages of planning my study, which made me even more leery. At some point, however, I needed to put aside my anxieties and begin exploring my options. I needed to trust my instincts and use my best judgment in terms of what seemed most feasible and productive. There always will be certain trade-offs and risks, but we need to think about which ones we are willing and most able to make, and which ones compromise our research the least.

needed to prove herself before they would trust her. She also needed to assure them that their time would be well spent and that it would result in something productive. This makes good sense when you think about it. We ask a lot of our participants in qualitative research, especially in terms of their time, so we should do everything we can to assure them that it will be well spent.

A different set of issues arises when researchers seek permission to use a K–12 classroom as a research site, especially if the researchers are teachers who wish to use their own classroom. Most schools have standard forms for parents to sign that give permission for their child's participation. A larger issue concerns how the teachers/researchers present themselves and the proposed research to parents. One approach is to use open houses at the start of the school year to inform the students' parents about the research. At these events, teachers can convey their desire to conduct research that will improve their teaching as well as the overall learning environment. Another approach is to write letters to families that outline the same information. What's important is for parents to be informed of the process; to understand the teacher's expertise and enthusiasm for doing the research; to be assured that high quality teaching and learning will not be compromised; and to be reassured that the teacher will be available at any time to answer questions and hear concerns. Helping parents feel comfortable with this process is vital.

Rapport, then, becomes a critical factor at this stage of the research process. A big part of negotiating access entails establishing rapport with the gatekeepers. Whether they like you, feel comfortable with you, trust you, and feel respected by you will all have an impact on their decisions at this stage. Being respectful, flexible, honest, responsible, and professional will help you establish a strong rapport. And when you move beyond the gatekeepers to the actual participants, rapport will also be important. Whether you are researching a setting that you are part of already (i.e., teachers in their own classrooms; employees in their own workplaces) or researching one that you are entering as an outsider, establishing and maintaining a strong rapport, both with the gatekeepers and with your participants, is essential.

There clearly are a lot of practical issues in choosing and negotiating entry into a setting. The more carefully you plan and negotiate this portion of your research, the better it will turn out. The setting you study and the people you ask to participate in your study will play an important role in your research.

Prompt 8: *Identifying a Setting.*

Think about the setting you would like to research. Why is it a good setting for your research question? What are some things you think you will need to do to negotiate entry into the setting? Consider the following:

- With and from whom will you need to talk and seek approvals? If you are not sure, whom can you ask?
- What concerns are those individuals likely to have, and how will you address those concerns?
- How will you establish rapport with these individuals? What can you do to facilitate establishing a good rapport with them?
- What obstacles do you think you might encounter in negotiating entry into the site and seeking approvals?
- How will you address those obstacles? What steps can you take in advance that will help you alleviate or minimize them?

Obtaining Human Subjects Approval

Whenever you do research with human participants, you are required, usually by your institution or employer, to obtain permission for that research. The primary purpose of obtaining this permission is to safeguard your participants. Human subjects review has a long history that dates back to just after World War II. At that time, numerous physicians, scientists, and other so-called professionals were tried at Nuremberg for their atrocious crimes against fellow human beings during the war. (For a thorough account of the history of human subjects review, see Paul Anderson's work in *CCC* in 1998 and in *Ethics and Representation in Qualitative Studies of Literacy* in 1996.)

Although many schools and employers have permission forms specific to their settings, you probably will still need to have your research reviewed through the university if you are completing it to fulfill an academic requirement. Now, colleges and universities that receive any kind of government funding, which are the majority of these institutions, are required to have institutional review boards (IRBs). These boards are comprised of faculty from across the disciplines who make judgments about the suitability of proposed research from the standpoint of human participants. Anyone whose research involves humans, regardless of the field in which the research is being conducted, must submit a human subjects proposal prior to carrying out the research. Further, because of recent, stricter guidelines, IRBs also often require formal training sessions for all faculty and students who carry out this type of research (you should check to see if your institution offers this training). The government takes research with human participants seriously, and it enforces the rules associated with it, which are aimed, first and foremost, at protecting our research participants.

Fortunately, research on writing typically will not pose any direct physical harm to or threaten human participants. However, there are ways in which writing researchers might act, or approaches they may take in collecting data, that are not in the best interest of participants. For example, researchers may act in ways that threaten the anonymity of participants or that expose or humiliate participants. Researchers who have some kind of authority over the participants in their studies (teachers over students, bosses over employees) may consciously or unconsciously connect participation with praise, better grades, or promotion. IRBs help prevent this. Therefore, you should begin learning about your institution's human subjects' requirements as soon as you know that your research will involve human participants. If you plan to carry out research in a professional setting—for example, in your school or in a workplace setting—then you should also check the requirements and obtain approvals from the appropriate individuals in those settings.

If you are a student, and if you plan to carry out research in a setting outside of your university, then you may still need the university's permission. It has been our experience that if you plan to disseminate your work beyond the university, then your project needs to be reviewed. If your work is intended only to add to your own personal knowledge and you do not intend to disseminate it, then it may not need approval, or it may be exempt. Specifically, certain research is deemed to be exempt because it does not pose any threat to human participants. However, it is still best to let your IRB make this determination. Also, be sure to leave yourself enough time for the review process. Some IRBs take up to a few months to review proposals.

Most often, IRBs request the following information from researchers to ascertain the ethical character of the research:

- The name of the investigator.
- The title of the project.
- The funding sources for the project.
- The nature of the project (whether it is new, a modification of an existing project, or a renewal) and its length.
- The numbers and types of participants.
- The identification of any special classes or groups of participants.
- The procedures for recruiting and informing participants of research procedures.
- A sample consent form.
- A list of any risks involved.
- A statement of how confidentiality will be maintained, including procedures for handling, storing, and disposing of data.
- An indication of the means by which research results will be disseminated and by which participants will be informed of the results, if they are to be informed.
- A list of any benefits to the participants.

IRBs also ask for copies of the research proposal and any instruments or tests that will be used in the research. Appendix C contains a sample application form and proposal for human subjects approval, including the consent form that was submitted.

Despite the rather bureaucratic feel that IRBs may give to your research studies, being ethical in your research need not be difficult. It usually is simply a matter of placing your research participants first. You always need to think about your participants, about how they perceive your work, about what their experiences are as participants, and about what their concerns might be at various stages in your research. The key aspects of ethical research are the following:

- Informed consent—your participants need to be informed about and freely consent to be involved in your research.
- Right to privacy—your participants need to be guaranteed privacy and anonymity if they desire it.

- Protection from harm—your participants should never be put in any situation that would cause them physical or psychological harm.
- Permission—your participants need to grant permission to you to use their work or statements as examples in your research (we contend that this extends to all situations where you use the work of others in any manner—e.g., even when you use students' assignments as models in subsequent classes).
- Method of use—your participants need to be reassured that you will use their statements and writing only in ways that are fair and ethical.

Again, on many levels, these requirements are quite basic; however, any research that involves humans, just like any human relationships, can become complex, often in unexpected ways. We address some of these ways and offer suggestions for handling them in subsequent chapters.

LENSES

The decisions you make as you plan your research—decisions about what you will do, where you will do it, and with whom—all will be influenced by both personal and theoretical perspectives. It's important, therefore, to make these perspectives explicit and to be aware of them as you plan your research. You should also be aware of how these perspectives can change as you plan your research. For example, as you review the literature, you are likely to encounter new theories that you end up using in your work. And when you begin considering different sites for your research and interacting with the gatekeepers and potential participants at those sites, you are likely to acquire new ways of thinking about and approaching what you're doing. You will start developing relationships with these individuals, and these relationships may change how you are viewing and approaching your research.

Prompt 9: *Thinking About Lenses.*

Reflect on and respond to the following questions, which are adapted from questions posed in one of the prompts in chapter 2:

- What in your own personal experience has influenced your thinking about how you will approach your research? In what specific ways are these experiences influencing your research?
- How are your *personal* beliefs and preferences affecting what you are planning on doing for your research, and where and with whom you are planning on doing it? What are those beliefs and preferences?
- How are your *theoretical* beliefs and preferences affecting what you are planning on doing for your research, and where and with whom you are planning on doing it? What are those beliefs and preferences?
- What are some biases you think you have that might influence your research?

We can best illustrate the impact of personal and theoretical perspectives on the planning stage of the research process through personal testimonies. One of our students, Mary Lou Wolfe, became interested in childhood development and early literacy experiences because of her background in science, her experiences with certain human services organizations, and her geographical location (she lived in a socioeconomically depressed suburb of Detroit). These factors influenced both what she chose to study as well as how, as her testimony indicates:

> I was filled with a sense of nostalgia yesterday when I heard that Mildred Benson, the author of the Nancy Drew stories, died. I remembered how much I enjoyed reading those books as a young girl. Suddenly as I reflected, two of the most moving perspectives that prompted my research—the joy of reading and the sense of mystery—surfaced. The early exposure to reading was my godsend. As a slow learner, I may never have achieved the level of success in my education if it were not for the ability to read and reread. I do not process input very quickly, especially auditory instruction, and therefore need to read and write things down in various ways to grasp them. This, of course, stimulates my imagination and hence the beautiful joy of reading, the wonderful excursions to imagined and real places. I have always strived to share the gift of reading freely with others, especially young children.

> The observation that people learn differently and personal exposure to the effects of illnesses (Thalassemia, rheumatic fever, and cancer), my own and those of other family members, led me to search for explanations in science and medicine. Although these disciplines were interesting, I found them mechanical and would rather explore the stories of people rather than physical processes. The theoretical perspectives of how we know and what we know, the fact that we can create our own reality, and that as communicators we can shape human knowledge were driving influences prompting my research. As a mother, it was a natural attraction then when I learned that Kiwanis, an international service organization, endorses the lives of young children as its first priority and that they strongly promote literacy at the earliest stages of development. With this knowledge, I joined the organization and plotted the plan for my research. …

> Reading to young children miraculously and mysteriously shapes their brain architecture, stimulates their knowing and use of language, and influences the course of lives. It became my pursuit to work within Kiwanis and with other service sectors in my Downriver community to increase awareness of the dramatic role that reading plays in the mystery of our knowing. (personal communication, 25 October 2004)

Another student, Diane Benton, became interested in developing a multicultural curriculum for composition classrooms. Diane is an African American woman who has taught composition and basic writing at community colleges, at a large state university, and even at an exclusive private middle and secondary school. Her positioning as an African American teacher, and her experiences teaching composition and literature classes to diverse student populations, had a profound impact on how she planned her project:

I am interested in language and African American culture. A very complex topic, I found that most of the texts and discussions in my graduate classes focused on the issue of literacy or the deficits of literacy in African American culture. Few focused of the strength of African American literacy or rhetoric. As I took classes and, at the same time, tried to determine my research topic, I kept hearing an inner voice that drew from my own experience as an African American woman and teacher who taught composition and African American literature. The voice kept raising the question of the unrecognized strength of African American rhetoric as was evidenced in the literature that I was teaching.

I also knew that there was not a lot of research that examined African American rhetoric as the center of the discussion. Few of my professors and peers, although encouraging, specialized in this area. I knew it would be easier to focus on literacy. In the end, I followed my inner voice. Though it required more time on an extensive literature review, it was the opportunity to answer questions raised in my own work and experience as both a composition and an African American literature teacher. (personal communication, 14 January 2005)

Finally, Diane Pons was strongly influenced by her experiences raising two children with chronic medical conditions:

As I raised two children with chronic illnesses, I became sensitive to the ways writing and medicine were intertwined. I depended on written texts for my medical education. The material provided the needed medical information, but it was so difficult that it left me wondering why there wasn't more medical literature created for patients. Most of what I read failed to present the patient's voice or address the emotional dimension of the medical condition. After my son's third major surgery, the relationship between writing and medicine expanded into an entirely new area as I began journaling. The simple act of writing in spiral notebooks made my stress more manageable, and I began to wonder about the therapeutic use of writing. My personal writing practice eventually led me back to school to work on a graduate degree in written communication.

The relationship between writing and medicine is at the core of all my research wonderings. Looking at medicine through my lenses suggested that the patient was someone who needed a stronger voice, the information to make informed decisions, and someone who needed to be involved in his/her treatment. My research question needed to address this proactive patient. My next consideration was the different ways writing would impact such a patient. Again, my personal bias popped up as I returned to my strong belief that writing had a therapeutic value. These lenses left no room for objectivity, so I knew I could not prove or disprove the therapeutic value of writing. Instead, I needed a research question that built on my experience.

After keeping a personal journal for years, I searched for ways to give my writing a social context. I began participating in workshops. I also began to consider questions that moved beyond "Is writing therapeutic?" to "What are the advantages or disadvantages of different genres?" and "How would one facilitate a writing workshop that encouraged healing?" These questions allowed for my biases and presented an opportunity to research more specific considerations of therapeutic writing. (personal communication, 17 July 2004)

As you think about your research and how you will approach it, you are likely to become attached to certain perspectives. Again, it helps to identify and be aware of these. They can influence your plans in a productive manner, shedding light on something new or helping you consider something in your research from a different perspective. They can also be limiting: You may end up holding certain perspectives at the expense of others, or you may become biased if you hold them too strongly. For example, if something you read really moves you, you might begin looking at what you are doing exclusively through the theoretical lens offered by that reading. Kenneth Burke's (1968) notion of terministic screens is useful here. We all possess certain terministic screens that are influenced by any number of factors and that, in turn, influence what we see and how we see it. What's important in qualitative research is that we be aware of these screens and that we keep an open mind about considering alternative ones. In other words, we need to acknowledge the lenses we bring to what we do, but we also need to be careful to not let those lenses blind us to other possibilities, especially in the early stages of our research.

Finally, you should also be aware of how your interactions with gatekeepers and/or potential participants might influence your personal and theoretical perspectives. If you talk to a gatekeeper who offends you off in some manner (e.g., you take a personal disliking to the person or that individual expresses viewpoints with which you disagree), you may, by extension, make certain judgments about the research site that may or may not be accurate. Conversely, if you really like those with whom you interact because their viewpoints are the same as yours, your judgments may likewise be clouded. You should avoid basing your decision solely on personal preferences and try to take other factors into account. Such preferences are important, but they should not be your only considerations. In fact, the nature of your study may well be such that your participants will not share your personal viewpoints. Again, you need to reflect continually on the perspectives you hold and on how those perspectives may end up influencing your research.

ETHICS

As you are planning your research, the key ethical concern you will have is with human subjects, which entails more than just an approval process. Underlying the official process should be a concern on your part, as the researcher, for the well-being of your research participants. From our point of view, being an ethical researcher involves being responsible and caring, and being concerned with the safety and well-being of the individuals who participate in your research. On the surface, this goal may seem fairly straightforward, and fortunately it often is. You likely know how to be honest and fair in your research and how to treat your participants with respect. In fact, just about any source you read that addresses ethics in qualitative research will stress that, as a researcher, you should always be responsible, first and foremost, to your participants. Concern for the settings you research should also be primary. Only then should you be concerned with yourself as a researcher and with your study. Again, there's a disposition involved here, and both of us believe that it is easier to acquire and maintain an ethical disposition if you begin to think about these issues as early as possible in the research process, even though time and other pressures may sometimes make it difficult to do so. (You may be worried, for example, that human subjects approval may delay your project or that such considerations are too time consuming. These kinds of worries are common, but they certainly should not supersede ethical concerns.)

Several of the questions that are important during the planning stage of your research have ethical implications, including the following:

- What setting will you research? (Different settings pose different ethical considerations. For example, schools tend to require various kinds of permissions, businesses may have concerns with proprietary information and confidentiality, social service agencies have concerns with confidentiality, etc.)
- Whom will you research? (Different participants may also suggest various ethical considerations. Are you carrying out research with children? If so, their parents need to sign consent forms. Are you researching individuals who might be considered part of a vulnerable population—for example, children, disabled individuals, individuals with special needs? If so, you need to take various measures to protect them. These are just a few examples of the considerations posed by your choice of research participants.)

Some other questions you should consider at this stage of your research include the following:

- How will you conceive of the individuals you are researching? What will be your position in relation to them? Will you be researching people who have greater or less authority and power than you do? (Both situations pose challenges; however, when you research those with less power and authority, you need to take special care so that you do not harm or coerce them in any way.)
- How will you interact with and treat your participants? (Will you interact with and treat them respectfully and fairly, especially if you are in a position of authority over them; e.g., you are responsible for their employee performance evaluation? Will you take precautions against deceiving or misleading them? Will you offer them something in return for their help; e.g., if you are a teacher giving grades, will you offer extra credit to students who agree to be interviewed? Again, there are numerous questions that should be asked concerning your interactions with your participants.)
- Whom will your research question influence most directly? How will your participants view your question? What investment will they have in it? (You may well have a greater interest in your research question than your participants. You need to be concerned, then, with how you approach them and engage them in your work.)

Considering all of these various questions will help you to establish an ethical stance in your research.

CONCLUSIONS

Often we hear our students, and even our colleagues, say they just want to get started on their research. They perceive planning as something that holds them up. We would contend that planning is a necessary part of research. You actually do get started on your research when you plan: You make key decisions that will influence what you will learn from your research, you lay out a road map that will guide you in your research, and you even begin exploring your question.

Of course, planning, like every other part of the research process, is not a discreet stage. You do not simply stop planning one day and then start researching. The choices you make as you plan are choices that you will continue making throughout the research process. And if things don't go just as you hope in your research, then you will likely need to go back and plan all over. Planning is productive work, but it is also challenging and ongoing work. Once you're in your research site, for example, you might discover that there are other individuals who can shed light on what you are investigating. As a result, you will seek permission to talk to those individuals. Or, if you discover documents later in your research that suddenly seem useful, you will likely ask permission to review them.

Research is a dynamic process. The best way to get a good start on that process is to plan it carefully. Using the tools and strategies discussed in this chapter will help you to do that. And you will want to keep using these tools while carrying out your research. For example, you should continue reading the literature, and you should keep an open mind so you can anticipate and then change those things you need to change in your research. Seldom does everything go exactly as planned in research, so be ready should the need for change arise. Effective planning will facilitate that readiness. It will also go a long way toward making your research productive.

APPENDIX A: SAMPLE PROBLEM STATEMENTS

Karen Reed-Nordwall, November 2000

One small area of writing seems to cause a major blemish on secondary writing curriculum. One area so tiny it's often ignored by teachers who feel it is not important, not real writing, not worth teaching in the small amount of time they have with kids. The blemish is spelling. While only a miniscule piece of the writing pie, it is still the one glaring obvious sign to parents, administrators, and teachers of a child's writing ability.

I have to admit I was one of the above teachers. I didn't know how to teach spelling, didn't feel it was important with computers that can check the paper for errors, but, most of all, I didn't know how to teach it. When a student came up to me to ask how to spell a word, I referred them to a dictionary. I just think of how many teachable moments went down the drain because of my assumptions and fears. Students need daily sustained reading and writing experiences, and I hoped this would be enough for them to become spellers but it is not. Spelling instruction needs to be intertwined into those experiences. The fact is, my students want to spell better. They are willing and ready. They just ask that I don't give weekly spelling tests. I decided I would try.

As a middle school teacher, I am concerned with helping my students develop all their literacy skills. While spelling is only a minor portion of writing, it tends to be separated in English classes and tested rather than taught. I find my non-writing students base their assumptions of their writing ability on their spelling ability. There are strategies that were developed over 20 years ago to help students with spelling, yet rarely do I see any of these strategies in place in the common Language Arts classroom. Still, I feel now may be the time to push spelling strategies since new programs are blossoming across the country, such as reading recovery and writing workshops. The "new" strategies permit teachers to teach spelling in relation to student writing and see improvement in students' self-esteem as writers.

Spelling instruction is important and should receive attention in the secondary writing workshop classroom. Spelling can fit within the reading/writing workshop philosophy, which centers itself around a few main principles: choice, response, and time. Students should be allowed to choose their spelling words, receive response from their teachers and other classmates, and have time in class to practice their spelling.

Most workshop teachers find themselves ignoring spelling all together or resorting back to drill and kill methods instead of teaching it. There must be other strategies that will help secondary students learn how to spell words. Rebecca Sipe discusses in multiple articles how testing spelling without teaching it and rote memorization are not effective methods for poor spellers. Constance Weaver urges teachers to help students notice spelling patterns for themselves. She asks educators to teach spelling within the student's writing—assess student drafts and use the mistakes to inform their practice. Margaret Hughes and Dennis Searle admit spelling is controlled by many factors but that teachers can examine misspellings and work with students to find strategies to help them learn how to spell.

Spelling is one of the major focuses of criticism against workshop classrooms, but if these teachers could think of strategies and avenues for incorporating spelling into their workshops without making it the focus, this could help strengthen workshop instruction to the community while helping students' literate lives.

Problem:

How can I teach spelling to middle school students in ways that will help increase their spelling awareness and locus of control?

Details:

Over the past year I have observed poor spellers, interviewed them, conducted visual memory tests, and examined drafts of their writing. I distinguished between which words they struggled with and investigated their backgrounds in spelling.

This year, I will target four students to observe their possible growth in spelling. These will be very poor spellers who I will locate based on a spelling assessment and first draft writing. I will then put some strategies into place for the whole class, small groups, and teacher/student conferences to help broaden spelling awareness and locus of control. Students need daily sustained reading and writing experiences to help them become competent spellers, but there are also strategies that can help them along. Some of the areas I would like to include in my spelling program are:

- Language should make sense and students need to see the relationships between words to help them spell words within families correctly. If you know how to spell one word, that can help you spell many more.

  — study eponyms, words from other languages, possessive apostrophe, word families, high-use spelling rules, and less common derivations, suffixes, prefixes

- Students need to know spelling strategies for how to spell words while they are writing and studying them.

  —have a go
  —see words everywhere (word walls); look for word in your environment to check it
  —study high frequency words, homophones, contractions
  —make own word list students can refer to
  —proofreading, drafting, peer conferencing
  —spell check on computers and personal spellers
  —take risks
  —play with language
  —look, say, name, cover, write, check
  —read and write in many different genres and for many different purposes

- Students need to reflect on spelling progress often.

  —reflect on why spelling is important (it should be because it helps to communicate your thoughts and not just be for teaching)
  —reflect on their spelling progress in their portfolios

I will observe the whole class progress, but especially my four target students, over the semester. Working one-on-one with my target students, I will be asking questions, such as why do they choose to write a word a certain way? I will ask them to take me through their thinking.

My students and I will examine their spelling patterns and try different strategies to help them with their spelling.

Methodology:

This will include observation, discussion/interviews, writing samples, and surveys. The focus will be on spelling awareness before and after the mini lessons and which strategies were beneficial to middle school students. Surveys will be taken before and after the spelling instruction and I will be assessing the students' drafts to see if they are incorporating the spelling strategies.

Lisa Tallman, November 2000

Problem:

The increased use of the World Wide Web provides consumer product manufacturers with another medium to provide their customers with information. However, the recent boom in online purchasing raises issues with regards to the type of information consumers can access and its presentation on a Web site, particularly in regards to safety information. Manufacturers must decide if safety information, such as product warning labels and safety information found on product packaging and in owner's manuals, should be placed on their Web site and how it should be designed.

The placement and design of safety information is an important issue to raise in light of current government and manufacturer actions and potential actions of consumers and the court system. For example, government actions, particularly in regards to health information, may set a precedent for providing safety information.

To address this issue, manufacturers will have to determine customer expectations, needs, and desires regarding accessing safety information. Based on this assessment, manufacturers will have to design safety information on their Web sites that meets customer needs and company goals, and that protects them from product liability.

Research Question:

Based on identified consumer expectations, needs and desires, what are the design guidelines for incorporating consumer product safety information on a Web site?

Background:

As an employee at an engineering consulting firm that specialized in the development and evaluation of product warning information, I became aware of manufacturers' concern with providing safety information on a Web site. Manufacturers wanted to know if they should provide safety information on their Web sites, and, if they did, how safety information should be presented. To begin to address these inquiries, my company conducted a survey of 30 Web sites to determine the presentation of Web-based safety information. (The results of the survey and a discussion of its implications were published in an article entitled, "Product Safety Information on the Web: Current Practices and Issues to Consider," *Safety News,* June 2000, pp. 1–2.) What we found was that few Web sites currently provided safety information and, among those that did, there was a wide variation on the type and presentation of safety information.

Further research shows the government is beginning to grapple with this issue, particularly in regards to medicines, but also consumer products. The Food and Drug Administration and the Consumer Product Safety Commission have solicited information from various industries, provided some initial guidelines, and praised organization's actions in regard to the presentation of safety information on the Web. The actions of these and other government agencies may set a precedent for the presentation of safety information and, eventually, lead to regulations requiring and/or governing it.

An initial literature review has shown that the issue of presenting consumer product safety information on the Web has received little attention. Consequently, a discussion of consumer expectations, needs, and desires regarding the presentation of safety information on the Web and how these can be incorporated into the design of Web-based safety information is almost nonexistent. This research seeks to address this gap.

Methodology:

Through a literature review, I will establish that this topic fills a gap that needs to be addressed. In addition, I will research government and legal actions that are setting a precedent for the placement and design of safety information on the Web and discuss their implications for manufacturers.

It will also be necessary to discuss the current state of the presentation of Web-based safety information. This will be drawn from, but also expand on, the survey completed through my employment. The process for this, perhaps informal survey will be to:

* Develop criteria for determining what types of consumer products are most likely to have safety information associated with them
* Determine a set number of products which consumers seek information for online
* Identify a set number of manufacturers that make these products and which have Web sites
* Review the Web sites for what type of safety information is presented and how it is incorporated into the site.

The focus of my research will be to conduct an online survey regarding what users' expectations are in regard to safety information when purchasing or attempting to find information on consumer products. This survey seeks to find out:

* What type of information consumers are searching for online
* What type of information consumers EXPECT to find on a Web site
* What type of information consumers WANT to find on a Web site
* Where consumers look to find information on a Web site.
* Whether consumers think manufacturers should provide safety information.
* Where consumers want safety information located on a Web site
* What format consumers want the safety information, i.e., PDF, HTML, other

Based on the results of this survey and research on developing Web sites from the technical communication industry, I will provide guidelines for the design of safety information on a Web site, considering issues such as format, navigation, and accessibility, and I will create prototype Web pages.

APPENDIX B: SAMPLE PROJECT PROPOSALS

Writing Project Proposal
Karen Reed-Nordwall
February 13, 2001

One small area of writing seems to cause a major blemish on secondary writing instruction. One area so tiny it's often ignored by teachers who feel it is not important, not real writing, not worth teaching in the small amount of time they have kids. The blemish is spelling. While only a miniscule piece of the writing pie, it is still the one glaring obvious sign to parents, administrators, and teachers of a child's writing ability.

I have to admit I was one of the above teachers. I knew that students needed daily, sustained reading and writing experiences, and I had hoped this would be enough for them to become spellers, but it was not. I didn't want to spend valuable time on an elementary concept. I didn't feel it was important with computers that can check the paper for errors, but most of all I didn't know how to teach it.

Spelling wasn't my first priority and I worried about how much time I would have to spend on it, but I began to see that many students who hated to write actually had poor spelling attitudes. It was then I realized how important this research project was. Students were being turned off to writing because of spelling. If I was going to convince students that their words are worth hearing, worth writing, and are powerful, I was going to have to begin with some basic skills, such as spelling.

When I began to look at my students' writing, I started to see a gap between what my students had learned about spelling and how they spelled in their writing. While they had received some instruction and a lot of spelling tests, they still could not spell in their writing. Their memorization of how words were spelled was not transferring to their writing. So how could I teach spelling so that it would improve their writing?

I have read many books about the process of writing and how to teach the process, but very few of these mentioned spelling. I have completely left spelling out of my classroom and now it is time to bring it back in. I have begun to ask myself, how can I teach spelling to middle school students in a workshop classroom that will help increase their spelling awareness and their self-esteem?

Making Spelling Instruction a Reality in My Classroom

My middle school classroom is set up as a reading and writing workshop based on the approach developed by Donald Graves, Donald Murray, Lucy Calkins, Nancie Atwell, and Linda Rief, to name a few. Workshops sprang from the Whole Language movement, which is a research-based philosophy of learning and teaching that rejects the belief that students need bits and pieces of language before they can become literate beings. Whole Language teachers try to create an authentic literate environment for students so students can incorporate these literate practices into their own lives.

Workshop philosophy centers itself on a few main principles: choice, response, and time. It is important that teachers continue to use their class periods to model how real readers and writers work. However, students still need instruction and this is why the minilesson is crucial. Workshop classrooms are student-centered; therefore, teachers can teach spelling through ten-minute minilessons and give the students the rest of the hour to apply the lesson to their writing.

I have been researching spelling now for almost two years and have begun teaching some spelling strategies to my students. The success has not been immediate, but I can see my students using some of the strategies and improving their writing. This encourages me to find out more about these strategies and how to implement them. I am most interested in how I can integrate spelling into my classroom.

For my Master's Writing Project, I would like to explore what strategies to teach during these ten-minute minilessons to help my students become better spellers. I would also like to develop other, longer spelling activities that might help improve my students' writing. I would like to design some minilessons, teach them, and observe four students to see if the lessons impact their writing. I hope to observe growth in my students' spelling ability and in my own ability as a writing teacher. As a practical outcome of this work, I hope to create a presentation that addresses how to teach spelling in a middle school Language Arts Workshop classroom. I will give this presentation to the teachers in my district at a conference this August.

Literature Review

When I began reading published works about how to teach spelling, a few key points kept reoccurring: the process of learning to spell is long and complex, there are strategies to help spellers, spelling should not be separated from writing because spelling is a writing skill, and spelling instruction should not be done if it does not affect the student's writing.

Spelling as a Long and Complex Process

In reviewing the literature on spelling, I first came across Diane Snowball and Faye Bolton's collaborative work. These authors speak mostly about spelling in the primary grades, but I found that many of their ideas and activities apply to older students too. In *Ideas for Spelling*, these authors stress how complex learning to spell is for children: "Spelling is a highly complex task that is gradually mastered over a period of time as an individual becomes acquainted with the properties and purposes of written language" (2). Students need time to try to create meaning; only when we teach kids about writing for an audience should spelling become a focus. These authors explain how many adults forget how complex spelling is for kids. We think that if kids spell it right on the test, they should be able to spell it right for the rest of their lives, and this is not true. When kids are writing, they are concentrating on making meaning and cannot possibly have everything perfect the first time through.

Bolton and Snowball believe that we cannot expect students to spell each word perfectly the first time. We are asking kids to think about an awful lot at once: lead sentence, details, complete sentences, writing about important topics, spelling every word correctly, using the right word, using strong verbs, including dialogue, thoughtshots, exploded moments, and reflective endings. That is an incredible list of things to consider. Writing is hard enough for an adult let alone an eleven-year-old. These authors plead with teachers to recognize the skills our students have and to celebrate those skills instead of concentrating on the highly visible and easily gradable spelling skill.

It is also crucial to remember the self-esteem of the writer. Two other authors, Laminack and Wood, discuss how students perceive themselves as writers based on spelling ability: "Many poor spellers have low self-esteem in relation to their writing ability" (*Ideas for Spelling* 2). Students are often taught that their words don't matter if they can't spell them. We need to remember that every time a student writes she is taking a risk; we need to nurture those risks or else we will stifle our students' growth.

Through these readings, I have begun to see that learning to spell may be strategic and that there are strategies teachers can use to help students learn to spell.

Strategies to Help Teach Spelling

The trend in education has been toward phonetic learning; however, Bolton and Snowball stress orthographic learning. The English language is complex if one studies phonemes, but it is simplified and consistent if one looks at meanings; therefore, these authors stress investigating morphemic relationships: "Orthography has focused on the units of meaning within words. When viewed in this way, the English orthography has emerged as being highly consistent and, in fact, near optimal" (*Ideas for Spelling* 4). Teachers have many avenues to teach meanings with compound words (if you can spell auto and mobile you can spell automobile), affixes, and word families. Students often see words as separate entities and feel relieved when they find out that words build on each other: if they know one, they know many more.

Bolton and Snowball look at competent spellers and use them as models to understand how one may become a competent speller. Teachers then have even more strategies at their disposal. The authors suggest these strategies for teaching spelling: word webs, reading and writing, drawing children's attention to print in their environment, adding high interest/high frequency words from readings to class word lists, conducting word searches from reading materials, identifying medial sounds, listening for words that contain a certain sound and listening for the number of syllables in a word, writing daily, and proofreading. They give examples of stories with discussion questions to help a teacher begin her search for how to teach spelling in her classroom.

Many teachers are concerned with how to assess student progress and give a grade, but these authors provide some innovative examples: observation, collecting writing samples, examining children's individual records, and spelling interviews. All of these ideas prove to me that there are many different approaches to teach and grade spelling and that there have been for years. So are my students only receiving one style that is decades old?

Other specific techniques appear in the July 2000 issue of *English Journal*. In this issue, Kelly Chandler explores how she wishes she would have taught spelling when she was a high school teacher. She gives some meaning-based minilessons that teach strategies to secondary students: "Mini lessons on the role of prefixes, suffixes, and roots in spelling will help students learn to spell far more than the words in question, as students who learn from them will be able to use that knowledge in other writing contexts, and in some cases, other content areas" (92). Helping students to generalize between words will help their spelling; for instance, if students can spell *compete* but misspell *competition* then you can help them go back to the base word and look at adding the suffix *-tion*. If they learn that rule they will be able to apply it to many other words. As long as the rules make sense and can be widely used, students will embrace them.

Chandler pushes teachers to help kids investigate words, patterns, and historical contexts and then invite students to find other examples of the rules and patterns. But her main point is to start discussing spelling with students: "Discussions of the words collected by the class can help kids connect what they know about spelling one word to a number of other words, as well as promote vocabulary development in an authentic way" (*English Journal* 92). We need to teach in a way that helps kids function as readers, writers, and thinkers but that also helps them in every curricular area. Writing has so many components. They all build on each other and cannot be separated, yet they must have a purpose, and this is where I see many spelling programs failing. They fail when they do not help students become better writers.

Spelling strategies will help students to grow into writers even past the time they walk out of our doors. In *Creating Support for Effective Literacy Education*, Weaver, Gillmeister-Krause, and Vento-Zogby discuss the importance of spelling strategies for developing writers: "Teaching children strategies for correcting spelling is far more important than giving them the correct spelling of any particular word" (*Creating Support for Effective Literacy Education*, Fact Sheets). The authors express the idea that students need to be independent spellers. Students often ask a teacher how to spell a word. Instead of giving them the answer for that one word, the teacher should give them strategies so that they can independently spell many other words as well.

Spelling Instruction Must Be Integrative and Must Affect Students' Writing

A month ago, I attended a Lucy Calkins seminar on how to teach reading. She discussed how some reading teachers have kids make mobiles about their books. She then asked why you would have kids make mobiles unless you wanted to see kids pull out their mobile during silent reading time and play with it? You wouldn't. Every activity should have a purpose for reading time and should help support student's independent reading. This is the same with writing. If we want our kids to spell well then we need to give them strategies that will help them during independent writing time.

Students need to study words and do spelling activities, but only if those activities affect their writing. In addition, we should be able to see that influence during writing workshop time. If it doesn't help writing, why teach spelling at all? Laminack and Wood discuss this issue in *Spelling In Use*: "Writers need tools (spelling, genre structures, punctuation, etc.) to write with power. The tools by themselves, though, are not important. Powerful writing is important" (113). Again, if it doesn't affect writing—make it better—why do we teach it? It's not just about spelling. This goes for all of our teaching. Whatever I teach, I want it to help with the main core of Language arts, which is reading and writing. Otherwise, it's a waste of time.

Students need to be in the position of writers to see why their spelling is important. It is important for communication and, therefore, we should help kids see their words as important and worth publishing: "For spelling instruction to make sense, it needs to happen alongside the real work of writers. This is the only way teachers can know what a child can do and what that child needs to know" (*Spelling in Use* 113). Real writers do not sit down and study ten words in isolation. They write and try out new words, make mistakes, and fix them. Kids should be able to have the same freedom to find their voice.

Many teachers are concerned with how to assess students' spelling without tests. They ask if they should give a writing grade and a separate spelling grade. This would further divide the skill of spelling from its counterpart of writing. Laminack and Wood quote Sandra Wilde to address this question:

> Sandra Wilde has noted that we feel compelled to give grades in spelling because it is a feature of writing that is very visible, even though it is not in any way the most important feature. She has said that giving separate grades in spelling and writing is like giving separate grades for multiplication and mathematics. (*Spelling in Use* 112)

Writing and spelling should not be separated. If the purpose is to make writing stronger, then improving spelling is one skill that must receive attention. This is similar to improving mathematics; if we are to improve our student's math, we must teach them multiplication. Math teachers would never consider separating multiplication from their math calculation grades, so why do English teachers assume that we can separate spelling from writing?

Moving From Questions to Decisions

The more research I do, the more I realize I will not find or be able to give any easy answers or quick fixes on how to improve spelling. I do find it helpful though to look at the questions Bolton and Snowball propose that may help us make good decisions about how we should teach it. These questions include the following:

- Am I dealing with spelling in relation to writing?
- Am I developing spellers willing to take risks, by providing positive reinforcement for each attempt?
- Am I aware of the children's self esteem and self-image?
- Am I assisting children to develop beyond their initial capabilities?
- Do the activities assist children to see relationships between words?
- Do the activities encourage children to form generalizations?
- Do the activities develop an interest in words?
- Are the activities really spelling activities?
  (*Ideas for Spelling* 97–100)

These questions led me to begin looking at the book that propelled me to try workshops in my classroom. I turned to Nancie Atwell's book *In the Middle* to see how she taught spelling in her class. This is a text I've turned to for so long now to help me teach reading and writing. In her writing workshops, kids have time to write and read in class, they have choice in what they read and write, and they receive feedback from the teacher.

Atwell begins with a reassuring message that learning how to teach spelling is a process that teachers go through: "I've moved from teaching spelling solely through students' editing and correcting of drafts to a three-pronged approach that includes more opportunities for exploration of patterns and rules, as well as weekly, individualized, word studies" (196). When she began to teach spelling, she taught it like I did, through helping kids edit their drafts. However, now she has moved into an approach that works best for her and her class: "My spelling minilessons take place on Tuesdays. Although I don't necessarily conduct a spelling lesson every Tuesday, it is always the day my students prepare individual word studies for the week. ... " (196). I like the idea of trying to push myself to concentrate on spelling once a week, but allowing for flexibility to meet students' needs.

Atwell teaches spelling through minilessons and has students choose words they want to study:

> In spelling minilessons I present information about the history of English, strategies of spellers, how to proofread, procedures for studying words, spelling resources, patterns, rules, and exceptions. As with all minilessons, I try to avoid instruction that isn't relevant to my kids. (196–197)

Atwell allows her kids to choose their words and teaches them valuable lessons about reading and writing. Her methods are always based on the current research with her kids' interests in mind. She teaches spelling by having students choose five words on Tuesday that they want to study and then they get in partners on Thursdays for spelling tests. She says: "I don't want to overload them or promote memorization-for-the-test; I do want to acknowledge that retaining spellings is hard and time-consuming work" (*In the Middle* 198). She allows her kids to choose their words from their writing and has partners team up to test each other on their words. Her approach allows students to have some choice on which words to study, and the words are from the student's writing. This way there is a direct correlation between writing and word spelling. I like the idea of spelling minilessons on patterns, but I do not think spelling tests are the answer for my students.

I knew how I did not want to teach spelling, but I was not sure how I did want to teach it. I used the questions from Bolton and Snowball to help me find a starting point for considering how to teach spelling in my workshop classroom. However, I knew I needed to turn from the books and begin to look in schools. (The annotated bibliography at the back of this document lists the published materials I reviewed.) I tried to find teachers who could help me answer these questions, and I found two college professors, two high school teachers, and one other middle school teacher. Together, we created a research group to try to understand how to teach spelling past elementary school. Each member of the research group examined four poor spellers and collected data. Together, through analysis of spelling autobiographies and draft errors, and through interviews, we found four categories of challenged spellers (See Figure 3.1).

Category 1

1. Strong readers and writers.
2. Strong internal locus of control.
3. Multiple self-correction strategies.
4. See spelling as secondary to meaning and correctable in editing.
5. Would strongly benefit from strategic instruction to address the t ypes of errors made.
6. May even have very low visual memory scores.

Category 2

1. Average readers and writers.
2. Some sense of locus of control.
3. A few self-correction strategies (multiple strategies) but generally rely on editors, the way words look, spell check, or dictionaries.
4. Very weak delayed visual memory.

Category 3

1. Reluctant readers and writers (weak).
2. External locus of control.
3. Recognize selves as poor spellers…little idea of what to do about it.
4. Associate lack of success with lack of interest, lack of effort.
5. Low visual memory scores on both short-term and delayed memory.

Category 4

1. Weak readers and writers.
2. External locus of control.
3. Out of touch with needs as spellers.
4. No sense of ability to influence the sit uation.
5. No strategies beyond sounding words out.

Figure 3.1: Categories of Challenged Spellers.

Now I wish to use the findings from this project and my readings to design mini-lessons that I can use in my middle school workshop classroom. There aren't books that tell how to do that. That is why this new research I'm proposing is crucial. I need to know how to teach spelling in a workshop environment where spelling is not the whole picture.

I will focus in my research on how I can teach spelling to middle school students in ways that will increase their spelling awareness and locus of control over language. I know I do not want kids to do spelling tests because I can see it does not work for most students. I do not know how to teach spelling though. Therefore, I wish to turn my attention in my research to how I can teach spelling in a productive manner.

Approaches

Since I teach in a workshop classroom, minilessons are a wonderful way for me to teach spelling strategies. I will designate Tuesdays for spelling minilessons each week. Since I want to spend only ten minutes or so on direct instruction and the majority of my time in a student-centered classroom, minilessons on spelling at the beginning of the hour will allow students to listen to the lesson and then directly apply it that day to their writing. In this way, spelling ideally will become intertwined with writing.

This year, like Atwell, I will start using some spelling minilessons in my classroom. The minilessons I have taught so far this year include: Roots, Portfolio Reflection and Discussions, and Proofreading. I will now begin implementing the final minilessons into my workshop classroom. Those include: Have a Go, Shape of Words, Word Wall/Homophones, Mnemonic Devices, High Use Spelling Rules, and Dictionary Skills. Some of these strategies help students while writing, while others raise students' awareness of problem words.

I need to incorporate the above minilessons into my teaching and then revisit them many times throughout the semester. I will do this because Snowball and Bolton caution teachers to choose a few strategies and then reteach them so as not to overwhelm the students.

I will implement the minilessons through whole class instruction with the hope of discovering a spelling program that will work in my workshop classroom. Also, I will survey the workshop teachers in my building about their own use of spelling in their classrooms, and about problems/concerns they have about teaching spelling. Since I will be designing a presentation for teachers, I need their input into what they need to know more about.

Assessing If It's Working

While I will be assessing the spelling awareness and growth of the whole class through their writing, I will target four poor spellers to observe their status in spelling over the semester. I want to study four students who struggle with spelling so that I can take a deep look at their spelling habits, attitudes, practices, errors, and possible growth.

I will choose these four poor spellers based on a spelling assessment and first draft writing. I want to see what problems they are having and what strategies they use. I also want to help them to use the strategies that I see competent spellers using. These will be described in the minilessons. I will then put these strategies into place for the whole class, for small groups, and during teacher/student conferences to help broaden my students' spelling awareness and locus of control over language. I will assess this program through an assessment of students' writing through collection of samples, through Portfolio Reflections, and through the dictation of early drafts. The focus will be on spelling awareness before and after the minilessons and which strategies are beneficial to middle school students who are poor spellers.

Assessment of Spelling

Richard Gentry designed a list of spelling words that students at certain grades should know. I had all of my students take the spelling assessment from fourth grade to sixth grade to see which grade they should be placed in according to spelling. I gave my students this test in September to locate my challenged spellers, and I plan to give it again to my challenged spellers to see if their spelling improves once I have taught the minilessons and projects.

Collection of Drafts

After I have targeted my four poor spellers, I will begin looking at the first drafts of their writing and assessing their errors. I will save all their drafts and then compare them. After I have taught the strategies, I will then look at their final drafts to see if they have used the strategies and improved their spelling in their final draft writing. I have taught my students to try their best in first draft writing and to worry more about spelling in the third to final drafts; therefore, final drafts will provide good data to tell me if they are improving.

Portfolio Reflections/Discussion

My students and I will work together to examine their spelling patterns and to try different strategies to help them with their spelling. I will use editing conferences with my four poor spellers to discuss their spelling errors and ways to improve them by using the strategies I have taught them. Students will have a chance to reflect on their progress though their portfolio reflections every ten weeks. This will help me see which strategies they are using and if they are comprehending them. Also, students will be reflecting on their progress in spelling, which brings spelling to a conscious level and may help them feel that spelling is something they can control.

Dictation of Early Drafts

Another kind of data that I will collect will be from the students' early drafts. I will keep a draft from September and dictate it back to them in April and compare how they spell the words at this later date. I feel most comfortable with this approach because it makes use of their own words and is connected to writing. Also, I feel this will show me their own personal growth in spelling.

Writing Project Document

When the research is complete, I hope to develop a spelling presentation about how I teach spelling in a middle school classroom. I am on "The Process Writing Team" in my school district, and I am in charge of presenting to teachers about teaching writing. I have designed a six-hour presentation on the method of writing workshop and would like to have a follow-up mini-conference this August addressing some more precise concepts such as how to teach spelling in a workshop classroom, how to teach research writing in a workshop classroom, and how to teach other writing elements. I work with ten other teachers who are creating presentations on some of the above topics. I think a spelling presentation would be very worthwhile. It will give the elementary school teachers as well as the secondary teachers access to the current information about teaching spelling.

I want to present about what worked with my students and what did not. I want to begin by discussing my research and what the current published research has said about teaching spelling. I hope to outline the specific minilessons I used with my students and their reactions to those minilessons. I also hope to discuss my journey in this alternative spelling program. The presentation will begin with the research I did last year with my group but will then focus on what I did in my own classroom for this study.

Time Line

January: Choose four challenged spellers to participate in study over the next four months. Collect their drafts and use their spelling assessments to find their spelling grade level. Begin to analyze their drafts. Give surveys to teachers and have them returned or seek them out. Analyze survey data.

February: Implement remaining spelling minilessons and activities: shape of words/word wall, mnemonic devices, have a go, and high-use rules. Take notes in journal about what I observe during these lessons and throughout the month.

Begin writing project presentation-introduction to past research.

Observe the four students and the rest of the class. Are they using strategies? Analyze previous data.

March: Continue to analyze data and observe my class to see what strategies they are using when writing. Teacher-conference with students about spelling strategies.

Write about how the activities of roots, high use rules, and proofreading worked in my classroom. Obtain feedback about the first draft of the presentation introduction and layout.

April: Continue to analyze data and observe my class to see what strategies they are using when writing. Give the spelling assessment again to my four students to see if their spelling grade level has changed. Observe their drafts to see if there is an improvement.

Conference with students about spelling strategies. Take break. Continue writing. Talk with reader. Dictate their earlier paper to them and compare the difference.

Read the portfolio spelling reflections of my four challenged spellers to see if they are reflective on their spelling achievement and feel more in control of their spelling.

Write about the minilessons: Shape of words/word walls, have a go, and mnemonic devices. Make revisions to chapter. Talk with reader.

May: Complete presentation. Give to reader. Begin my reflection for Project and collect the other two papers for the Project.

June: Make revisions. Give to reader and begin revising again. Finish Writing Project.

Designing Safety Information for the Web: Allowing Consumers to Drive Design
Lisa Tallman
January 1, 2001

Problem

The increased use of the World Wide Web supplies consumer product manufacturers with another medium to transmit information to their customers. However, the recent boom of Internet use raises issues regarding the types of information consumers can access and its presentation on a Website, particularly in regards to safety information. Some companies selling consumer products online provide access to product literature, such as owner's manuals, repair manuals, and installation sheets. On the other hand, many companies do not furnish this type of information on their Websites. Instead, they direct customers to acquire product information through more traditional channels, such as phone or mail. With increased competition for Web consumers, manufacturers need to decide what role the Web plays in the dissemination of their product information. Manufacturers must ask themselves if the information they are supplying fulfills customer needs and expectations, along with the company's marketing and pubic relations objectives. To complicate matters, manufacturers also must honor any applicable government regulations to ensure product liability protection. Consequently, manufacturers must determine if safety information, such as that found on product packaging, in owner's manuals, and on product warning labels, should be placed on their Websites. If manufacturers do decide to put safety information on the Web, they then must determine how it should be presented in terms of design.

The placement and design of safety information are important issues to raise in light of current manufacturer and government actions. I had the opportunity to participate in a survey conducted by my former employer, an engineering consulting firm named Applied Safety and Ergonomics, Inc. This survey found that a few manufacturers do provide consumer product safety information on their Websites.

The survey also found that there is a large variation in how safety information is presented (Young, et al. 2). Manufacturers are not the only entities grappling with this issue. Government agencies are also aware that the Web can be a means of communicating safety information and are challenged with regulating it.

Government actions, particularly in regards to health information, may set a precedent for conveying safety information. For example, Byron Tart, Director of the Center for Devices and Radiological Health, stated that medical device information on the Internet "likely" constitutes labeling. This is significant because the Food and Drug Administration (FDA) is responsible for the regulation of product labeling. If information placed on the Internet is considered labeling, then it would fall under the jurisdiction of the FDA and its regulations. In response to this statement, the FDA issued a *Federal Register* notice announcing a public meeting to discuss issues related to the promotion of FDA-regulated medical products on the Internet. The FDA recognized that it needed input from interested and affected parties to begin making decisions about regulating Internet content (Moberg, et al. 217).

In a related government action, the Consumer Product Safety Commission (CPSC) developed a "Top Ten List" of safety principles aimed at reducing the number of recalls. In developing this list, the CPSC supplied examples of manufacturers practicing these safety principles. In "The Product Safety Circle: Catalog of Good Practices," the CPSC lauded Toys 'R Us for adding "a useful safety innovation to its website, when it began adding age labeling guidance and safety warnings for products advertised on the site" (2). These are just two examples of the actions the government is taking to address the issue of Web-based safety information.

Michael Mazis and Louis Morris view the Web as a new and potentially effective means of communicating risk information. These authors state that "the Internet may be used to transfer rapidly large amounts of customized information and of 'high quality' information" (115). However, they place a caveat on using the Internet as a channel for communicating safety information. The Internet "will present opportunities and challenges for risk communicators. However, much remains to be learned about technology and about how consumers will use this technology" (117). Mazis and Morris clearly place the decisions for the provision of safety information on those who will use it—the consumers. Manufacturer and government actions should be driven by consumer preferences. For my thesis, I will set out to determine what needs, expectations, and desires consumers have for the presentation and availability of safety information on the Web. Using these consumer needs and preferences as a foundation, I will set out to develop design guidelines for incorporating consumer product safety information on a Website.

Background and Thesis Rationale

As an employee at Applied Safety and Ergonomics, Inc., I had the opportunity to participate in a survey to determine the current state of the presentation of Web-based safety information. This survey was conducted because of inquiries made by the firm's clients, manufacturers of consumer products. The engineering consulting firm specializes in the development and evaluation of product warning information in print, including owner's manuals and product labels. However, with the increasing use of the Web as a communication channel, the firm's clients began to inquire about whether they should supply safety information on their Websites and, if they did, how that safety information should be presented.

To answer this question, the firm conducted a survey of 30 Websites. This survey sought to assess the current state of "the prevalence and characteristics of product safety information" available on these sites. Young, et al. selected 10 classes of products and then selected one product from each class (see Table 3–1). For each product, three sites were chosen. The products were chosen because they represented products that (1) had potential hazards associated with them, (2) consumers might seek information for on the Web, and (3) could potentially be recalled (1–2). The authors admit their sample was clearly "too small and limited to permit much generalization about the 'state of the art' with regard to the Web-based presentation of safety information" (2). However, the authors believe that the sample was sufficiently diverse to provide some information on how manufacturers currently present safety information.

TABLE 3.1

Product Class	Product Category
Major household appliance	Refrigerator
Minor household appliance	Toaster
Personal care product	Hair dryer
Audio/visual equipment	Television
Entertainment accessory	TV stand
Household furniture	Recliner
Household cleaner	Floor cleaner
Outdoor use product	Dual-fuel stove
Power tool	Power drill
Over-the-counter medication	Pain reliever

This survey did uncover some information; however, it also produced more questions than clear-cut answers. This, in part, was due to the variety of ways Web-based safety information was presented. With no formal guidelines for format, navigation, or architecture, safety information that existed was presented in many forms and in several locations on the sites we investigated.

For example, safety information was presented in:

* HTML pages that contained identical information as the product packaging,
* downloadable (PDF) owner's manuals and material safety data sheets,
* links to product recall information,
* a warning located in warranty information, and
* Frequently Asked Questions sections of sites.

Yet, the survey found that few manufacturers furnished safety information or links to safety information on the actual product Web page.

This variety of presentation methods has led me to want to develop design guidelines for Web-based safety information using consumer needs, expectations, and desires as a foundation. Admittedly, manufacturers will have to take into account more than just consumer needs and preferences when supplying safety information on their Websites. Manufacturers have created Internet presences to remain competitive and increase sales. Consequently, information supplied on a Website must meet the sales and marketing goals of the manufacturer. In addition, Websites function as public relations and advertising tools. Product safety information can easily be seen as an obstacle to achieving these goals. Therefore, manufacturers must balance the information presented on their sites to achieve the desires of the company and, simultaneously, those of the consumer.

To complicate matters further, manufacturers must also fulfill their obligations under the law. In "The Law Relating to Warnings," M. Stuart Madden points out that "product manufacturers have a duty to provide instructions and warnings sufficient to permit a product to be used safely or to enable a user to make an informed choice not to use the product" (315). This duty to warn is dictated by government regulations. As mentioned previously, government agencies are tackling the issue of creating regulations to determine manufacturers' duty to warn on Websites. They have begun to develop guidelines for the presentation of Web-based safety information and have singled out examples of manufacturers setting good examples in this area.

In summary, there are a host of issues, goals, and perspectives that manufacturers must consider when presenting safety information on their sites. Sales, marketing, public relations, and advertising objectives of the company, obligations to follow government regulations in order to protect themselves from liability, in addition to consumer needs and desires, all come into play when manufacturers determine what information should be furnished on a Website.

So why use consumer needs, expectations, and desires as a foundation? Because "the Internet is user driven. The interaction is initiated by, and terminated by, the user" (Helander and Khalid 610). "It is therefore imperative to understand who the customers are and what they want" (Helander 2–770). Without the consumer, there would be no reason for manufacturers to have Websites. Without knowledge of consumer needs and desires, manufacturers cannot successfully achieve sales, marketing, public relations, or advertising goals. Without considering the consumer, manufacturers will not be able to fulfill consumer needs, expectations, and desires in regards to product information. If that information, in this case safety information, is hard to find, not clearly identified, or hard to read on the computer screen, it is useless. A study of product information seeking behavior can provide the information needed to ensure that consumers will find safety information through clearly identified navigational tools and be able to read and comprehend it. Consumers are, obviously, the best and most likely place to find data on what consumers need, expect, and want.

To design Web-based safety information for consumers, a means of getting information from them must be created. My thesis provides that means. By surveying consumers, I will be able to develop guidelines that will be beneficial to their search for product information. As a result, manufacturers will be able to present safety information so that it meets consumer preferences. At the same time, manufacturers can fulfill their obligation and desire to provide their customers with the proper safety information at the right time while ensuring that company goals are achieved and protecting themselves against product liability.

The above discussion begs the question: Do consumers want safety information provided on a Website? In "What's Driving Shoppers Away?" an article on the failure of e-commerce sites to get consumers to actually purchase a product, Carolyn Snyder argues that "most products have a downside or fine print, and users want to know the full story before they decide to buy. By the time users show up on the site, they're looking for more depth than they get in a 30-second TV commercial. ... [U]sers indeed want balanced information. ... " (3). Safety information would certainly provide the balance consumers are seeking.

Additionally, whether consumers "want" safety information, is really a secondary consideration to whether consumers "need" safety information. Madden states: "A claim that a manufacturer failed to provide adequate warnings or instructions is probably the most prevalent element in modern products liability litigation" (316). Clearly, consumers have shown through their use of the legal system that safety information is important to their decision to purchase and use a consumer product. Fully informed consumers need to have safety information provided whether they expect or want it. To be fully informed, consumers must know the hazards associated with consumer products and how to avoid them. To know this, consumers must be able to access and understand the safety information manufacturers are, in most cases, required to provide with their products. Ultimately, consumers must be informed to ensure their own safe use of a product. The Web can be an effective channel for communicating this information. Therefore, research must be conducted to identify consumer needs, expectations, and desires regarding the presentation of safety information on the Web so that consumers will get the safety information they clearly need.

My research has the potential to not only benefit consumers, but manufacturers, government agencies, and the technical communicators and Web designers responsible for creating Websites. Madden asserts:

> "All too frequently, however, products liability trials involving warnings issues have proceeded in blithe disregard of any scientific examination of how warnings have worked, or have failed to work, in a particular setting. Many manufacturers craft warnings without subjecting them to prior testing or trial for effectiveness, and indeed until very recently most warnings traced industry custom rather than any fresh and scientifically based knowledge of risk communication" (329).

My research will help manufacturers "craft" their warnings effectively for the Web.

As a technical communicator, I also seek to assist those in the technical communication and Web design industries who may be responsible for presenting safety information. It is a common complaint among technical communicators that we do not get to talk to our users enough. Through this research, I will be able to speak with and gain information from Web consumers.

If successful, my research may even be used to help government agencies, such as the FDA and CPSC, develop guidelines that can be disseminated and used by interested parties, such as manufacturers and the firms who work with manufacturers, in the development of Web-based safety information.

In conclusion, the consumer should be used as a foundation for determining the presentation of Web-based safety information. As Helander and Khalid declare, "the customer is of central interest" (612). Using consumers as a basis, the decisions manufacturers, government agencies, technical communicators, and Web de-

an example of this design shown on Merck Research Laboratories' Website. In Merck's Web advertisement for the prescription drug Zocor, when a consumer clicks on the word Zocor, product information describing warnings and necessary precautions appears on the screen (116).

Mazis and Morris' consideration of the issue of Web-based safety information lies not in whether to do or not to do it, but in the benefits of doing it and suggestions for how it should be done. However, Mazis and Morris admit that much remains to be learned about the Internet and how consumers will use it (117). Again, this provides a catalyst for the research I plan to conduct.

In the last article I would like to discuss, "Surfing the Net in Shallow Waters: Product Liability Concerns and Advertising on the Internet," Moberg, et al. speak specifically to liability concerns when putting safety information on the Web. The significance of this article lies in their summary of existing regulatory laws that may apply to Internet content. They also provide suggestions that manufacturers of medicines and medical devices can use to furnish safety information on the Internet and meet product liability considerations (214). The authors assert that the FDA is challenged to determine how the present regulations for advertising medicines and medical devices relate to the Internet as a channel for disseminating such information.

Moberg, et al. note that the FDA's very broad definitions as to what constitutes labeling and advertising, and regulations governing labeling and advertising, certainly place the Internet under its jurisdiction. However, this broadness complicates matters as to what does and does not apply to the Internet. Consequently, the authors argue that the Internet will become an "impetus for additional specific regulations" (216).

Moberg, et al. indicate that these "specific regulations" will not come to fruition without addressing several concerns, such as the presentation of product information, links between Websites, international issues, and other types of Internet media besides the home page, such as chat rooms and news groups. Each of these issues can only begin to be addressed when the consumer is consulted. The authors acknowledge the role of the consumer when they raise a significant question: "How should product information be presented to ensure Internet users will know that product information is available and where it is available?" (217). As stated previously, the Internet is user driven (Helander 2–770). Therefore, ultimate control rests in the consumers' hands. Government agencies should allow consumers to dictate, to some degree, what information should be furnished on a Website, how it should be presented, and where it should be located. To make any regulatory decision, the government must determine what the user needs and wants. My research addresses format, navigation, and accessibility issues from the consumer perspective.

Although each of these articles approaches the problem of safety information on the Web from the position of the product manufacturers and/or the government, they also show the importance of identifying consumers' needs, expectations, and desires on the issue of Web-based safety information. My research hopes to begin to identify consumer preferences and use these preferences as a basis for creating Web design guidelines for the presentation of safety information. The following section details the methods I will use to complete my research.

Methodology

To develop Web design guidelines for safety information from the standpoint of consumer preferences, I will employ several methods in my research: a literature review, online survey, focus groups, and usability testing of Web design guidelines and Web page prototypes.

Literature Review

My proposed research fits into the context of three areas of existing research: consumer use and expectations regarding the Internet, Web page content and design, and safety or risk communication. As noted previously, scholarly literature and empirical studies focusing on the presentation of safety information on the Web are almost nonexistent. But as my work with a firm that specializes in the development and evaluation of safety information shows, there exists a need, especially on the part of manufacturers, to address this lack of research.

As mentioned previously, my preliminary research has been limited to attempting to find literature on the combined subject of safety information and the Internet. As of yet, I have made only limited attempts to locate information on consumer Internet behavior, the presentation of safety information in other media, or current practices in regards to Web design. I will expand my literature review to encompass research undertaken in the following areas:

- Consumer opinion and advocacy groups, business, marketing, human factors, and technical communication professionals have researched consumer expectations regarding the use of the Internet for finding consumer product information and for consumer purchasing.

- Technical communication and human factors professionals have done extensive research on the communication of safety information in print.
- Government agencies have published documents that provide an explanation of the issue from a government perspective. These documents indicate the problems the government is having with tackling this issue. Additionally, these documents also provide preliminary guidelines for manufacturers in regards to putting safety information on the Web.
- Members of the legal community have published articles on the implications of the Internet on product liability.
- Industries such as information technology, business, technical communication, marketing, and public relations have developed Web design guidelines.

In addition to these areas, I will use the data collected in a survey conducted by my previous employer to document what manufacturers are currently doing in the presentation of their safety information. The survey looked at and recorded what safety information was found on 30 Websites. I will revisit these Websites to verify the documented results. Additionally, I will take a closer, more detailed look at formatting issues, such as whether safety information was provided as an HTML Web page or as a downloadable PDF in a manual or as a warning label. I will also look at navigation and accessibility issues, such as how the safety information is identified on the site and where it is located on the site.

In summary, my continued literature search will be focused on four areas: consumer use of the Web; consumer design preferences regarding format, navigation, and accessibility; current Web design guidelines; and manufacturer, government, and legal research and actions. I will continue to search for information in the technical communication industry, but expand my literature search to include the following disciplines: human factors, psychology, marketing, public relations, business, and computers and technology, as well as other disciplines that I may later identify. Also, I will continue to look for government and legal research and actions pertinent to this topic.

Online Survey

The primary goal of my thesis is to identify consumer needs, expectations, and desires regarding the types of safety information presented on a Website and how consumers would like it presented (i.e., format, navigation, and access). To achieve this goal, I will develop and conduct an online survey. I have chosen this method because its medium is the Internet. As such, it lends itself to reaching the population, Web consumers, from whom I need information.

Through the online survey, I will seek to identify factors regarding what product information consumers look for online, and their needs and expectations regarding safety information. At present, I plan to survey 500 Internet users. There are several companies, such as Insight Express, that provide surveying services. Additionally, several academic and nonprofit organizations also provide surveying services. After further research I will determine which service to use to administer a reliable, but inexpensive survey. There are several companies, such as CyberAtlas, Harris Interactive, and Media Matrix, that keep a check on the pulse of the Internet through various surveys. These companies constantly update demographics on the Internet population, such as the number of the US population online; the age, gender, and race of users; and income and education statistics. Since these organizations currently gather this type of demographic information, my survey will not seek to gather in-depth information on these topics. Instead my survey will focus on the following categories:

- The types of consumer products about which online purchasers seek information.
- Consumers' interest in accessing safety information.
- Consumers' previous experience with Web-based safety information.
- Consumers' opinions as to whether manufacturers should put safety information on the Web.
- The format in which safety information should be supplied (i.e., in the contents of a manual, on the HTML page containing product information and specifications, in the form of printable product labels, and/or downloadable as a PDF).
- Where consumers would search on a Website for safety information.
- Where consumers would want safety information located on a Website.

The online survey will be developed and shaped from information I gain through my research and from feedback from a small number of sample respondents and those interested in my work (i.e., my former employer).

Web Page Guidelines and Prototypes

This section of my research will be an iterative process that will result in the development of Web page guidelines that communicators of safety information can use to design Web-based safety information. Based on research from my literature review, particularly research on consumer preferences and Web design, and the data collected from the online survey, I will develop guidelines for the presentation of safety information on the Web and create prototype Web pages illustrating those guidelines.

For my prototypes, I will create Web pages that display safety information for a fictional product. I will present safety information on a Web page in HTML and downloadable in the form of a manual and/or product label. I will consider the issue of safety information presented in instructions that are found on Web pages. In considering format, I will also demonstrate the use of color, font size and type. The prototype Web pages will also illustrate where on a site consumers expect and want to find safety information and what navigational aids are needed to direct the consumer to that safety information. The prototypes will also show alternative ways safety information can be presented, such as through links and in floating windows. These Web pages will illustrate how safety information should be formatted based on consumer needs, expectations, and desires identified through the online survey and based on Web page design guidelines found in my literature review.

Focus Groups

After completing the Web page prototypes, I will hold two focus groups of 8 to 12 Internet users who look for product information online. The focus groups will view the Web page prototypes, and I will elicit feedback on the presentation of safety information, specifically how it is formatted, the navigational aids provided, and its location on the Website. Based on their feedback, I will reevaluate my guidelines and prototypes and incorporate changes.

TABLE 3.2

December 27–January 28	Complete literature review and update survey of existing Websites for the presentation of safety information.
	Deliverables:
	Thesis introduction, background, and literature review sections.
January 15–February 4	Develop online survey.
	Deliverables:
	Online survey.
	Thesis section related to the development of this component.
	Incorporate feedback on previous deliverables.
February 5–February 11	Obtain feedback on survey. Incorporate feedback into survey.
February 12–February 25	Conduct online survey.
	Deliverables:
	Begin draft of methods section of thesis.
	Update previous sections of thesis as necessary.
February 26–March 11	Analyze online survey results.

	Deliverables:
	Thesis section related to online survey results.
March 12–April 1	Create Web page guidelines and prototypes.
	Deliverables:
	Web page guidelines and prototype Web pages.
	Update methods section of thesis.
	Update previous sections of thesis as necessary.
April 2–April 8	Obtain feedback on Web page guidelines and prototypes. Incorporate feedback.
	Deliverables:
	Update methods section of thesis.
	Update previous sections of thesis as necessary.
April 9–April 22	Conduct focus group. Incorporate focus group feedback into Web page design guidelines and prototype pages.
	Deliverables:
	Thesis sections related to these components.
	Update previous sections of thesis as necessary.
April 23–May 6	Develop usability test method and related materials (questionnaire).
	Deliverables:
	Usability test methodology and related materials.
	Thesis sections related to these components.
	Update previous sections of thesis as necessary.
May 7–May 13	Obtain feedback on usability test method and materials and incorporate.
	Update previous sections of thesis as necessary.
May 14–May 27	Conduct usability testing and analyze results.
	Deliverables:
	Update thesis sections related to this component.
	Update previous sections of thesis as necessary.
May 28–June 10	Finalize Web page guidelines and prototype Web pages.
	Deliverables:
	Web page guidelines and prototype Web pages.
	Update thesis sections related to these components.
	Update previous sections of thesis as necessary.
June 11–18	Analyze and reflect on research project.
	Deliverables:
	Write thesis conclusion.
	Update and edit thesis in entirety.
June 18–July 6	Hand off final copy of thesis to director and reader. Meet and obtain feedback.
July 6–12	Complete final edits.
July 12	Mail to Graduate School.

Usability Testing

After incorporating the feedback from the focus groups, I will conduct two sets of usability testing on a second batch of prototype Web pages. In the first set, my respondents will be 5 to 10 Internet users who participated in the focus group. For the second set of usability testing, my participants will be 5 to 10 professionals selected from the following areas: technical communication, human factors, psychology, and Web design. At this time, I envision my usability test to take the form of a questionnaire to be used in conjunction with viewing my prototype Web pages. This questionnaire will seek to assess participants' reactions to the prototype Web pages I will develop and elicit comments, suggestions, and questions the respondents may have upon viewing and assessing the quality of the prototype pages. I will use this feedback to reevaluate my design guidelines, and to improve and finalize my prototype Web pages.

Works Cited

Consumer Product Safety Commission. "CPSC Chairman Ann Brown Unveils Product Safety Initiative 'Top Ten List' of Safety Principles Aimed at Reducing Number of Recalls." Press release. 5 Oct 2000. <http://www.cpsc.gov/cpscpub/prerel/prhtml01/01/003.html>.

Consumer Product Safety Commission. "The Product Safety Circle: Catalog of Good Practices." Undated. <http://www.cpsc.gov/businfo/pscgood.html>.

Consumer Product Safety Commission. "The Product Safety Circle Initiative." 2000. <http://www.cpsc.gov/businfo/psconcept.html>.

Hayhoe, G. F. (Ed.) "Heuristics for Web Communication. *Technical Communication* 47 (2000). Entire issue.

Helander, M.G. "Theories of Models of Electronic Commerce." *Proceedings of the IEA 2000/HFES 2000 Congress.* 2–770–2–773.

Helander, M.G. & Khalid H. M. "Modeling the Customer in Electronic Commerce." *Applied Ergonomics* 30 (2000): 609–619.

Madden, M.S. "The Law Relating to Warnings" *Warnings and Risk Communication.* Eds. Michael Wogalter, David DeJoy & Kenneth Laughery. London: Taylor & Francis. 1999. 315–330.

Mazis, M. B. & Morris, L. A. "Channel." *Warnings and Risk Communication.* Eds. Michael Wogalter, David DeJoy & Kenneth Laughery. London: Taylor & Francis. 1999. 99–121.

Moberg, M. A., Wood, J. W. & Dorfman H. L. "Surfing the Net in Shallow Waters: Product Liability Concerns and Advertising on the Internet." *Food and Drug Law Journal,* 53 (1998): 213–224.

Synder, C. "What's Driving Shoppers Away?" *User Interface Engineering's Eye for Design* 7 (2000): 2–3.

Young, S. L., Wisniewski, E. W., Tallman, L. A., Schiller, J. A., Frantz, J. P. & Rhoades, T. P. "Product Safety Information on the Web: Current Practices and Issues to Consider." *Safety News* June 2000: 1–2.

APPENDIX C: SAMPLE HUMAN SUBJECTS PROPOSAL

Eastern Michigan University
College of Arts and Sciences
Human Subjects Review Committee
College Application Form

Date Submitted: 1/23/03   Due Date of Funding Proposal: N/A

Principal Investigator: Karen Hoffman

Co-PI/Project Director: Linda Adler-Kassner

Department English/Written Communications Telephone: (734) 487 4220

E-mail:

Fax: (734) 483–9744

Title of Project: Place and Myth in an Ecologically/Locally Focused Curriculum

From what sources are funds expected for this project? None

I. Is this application: New

Will this project continue for more than one year: No
If this is a renewal:
Date of last approval by this committee:

_____

Principal Investigator in previous research:

_____

Describe any modifications to the previously approved research protocols. Were any human subjects problems encountered in the previous research? If yes, how were they handled?

II. If you are requesting an exemption from HSRC review, explain the statutory basis for the requested exemption.

III. Numbers, Types and Recruitment of Subjects

A. Numbers and characteristics of subjects (e.g., age ranges, sex, ethnic background, health status, handicapping conditions, etc.):

—Approximately 100 8th grade students at Emerson Middle School in Livonia, MI

—Ages 12 to 14 with a variety of academic abilities, including some special education students

—Primarily white, working class

—Approximately equal numbers of males/females

—Emerson Middle School is a Title I, low-income, at-risk school.

B. Special Classes. Explain the rationale for the use of special classes or subjects such as pregnant women, children, prisoners, mentally impaired, institutionalized, or others who are likely to be particularly vulnerable.

Little research exists on what motivates middle school students to take ownership for their own literacy; most research has focused on high school and college level students. Since middle school students have developmental differences from high school and college students, 8th graders are appropriate for the study. Since the curriculum meets the district requirements and is based on best practice in the classroom, the students will not be denied an appropriate education or harmed by the study.

C. How is the individual subject to be recruited for this research? Is it clear to the subjects that participation is voluntary and that they may withdraw at any time without any negative consequences?

I will explain to all of my classes the nature of my research project; I will make clear to them that it is for my master's project, that it is separate from class, the types of data that I may gather, that their names will not be used, and that their participation will not affect their grades in the class. This will be followed up with a consent letter (attached) for parents and students. Parents and/or students are free to request non-participation without penalty. The letter makes clear to students and parents that participation is voluntary.

IV. Informed Consent

A. To what extent and how are the subjects to be informed of the research procedures before their participation?

Again, I will inform students at the beginning of the semester about the research I will be doing for my master's project; I will make clear to them that they do not have to participate and that there will be no penalty or hard feelings if they choose not to participate. The consent letter further reiterates the purpose of the research and the nature of the proposed study.

B. A copy of the consent form is attached.

V. Risks Involved in the Research

A. Does the research involve any of the following procedures (YES / NO)

Deception of the subject: No

Punishment of the subject: No

Use of drugs in any form: No

Electric shock: No

Deliberate production of anxiety or stress: No

Materials commonly regarded as socially unacceptable: No

Use of radioisotopes: No

Use of chemicals: No

Drawing of blood: No

B. Any other procedure that might induce in the subject any altered state or condition potentially harmful to his/her personal welfare? No

C. Any procedure that might be considered an invasion of privacy? No

Disclosure of name or individual research subjects? No

Any other physically invasive procedure? No

If the answer of any of the above is "Yes", please explain this aspect of the research procedure in detail.

VI. Confidentiality

    A. To what extent is the information confidential and to what extent are provisions made so that subjects are not identified?

    The information provided by students will be used to analyze the effectiveness of an ecologically and locally based language arts curriculum on student motivation. Thus, in written reports about this subject, quotes or samples of work from them may be included. All identifying characteristics will be removed from quotes and/or student work samples. Pseudonyms will be assigned to all students.

    B. What are the procedures for handling and storing all data so that the confidentiality of the subjects is protected (particular attention should be given to the use of photographs, video and audio recordings)?

    Observation notes, samples of students work, and all other data will be stored in a locked closet file cabinet in my home in Milford. Last names of students will be removed from student work samples.

    C. How will the results of the research be disseminated? Will the subjects be informed of the results? Will confidentiality of subjects or organizations be protected in the dissemination? Explain.

    The research results will form the basis for at least one professional article and/or presentation analyzing student ownership of literacy. Subject confidentiality will be maintained throughout the dissemination. As above, pseudonyms will be used for each subject, and each time a subject is mentioned in any publication they will receive a new pseudonym. Students will be given a brief overview of the results at the end of the semester, but individual work samples or quotes will not be shared.

VII. Describe any anticipated benefits to subjects from participation in this research.

    —Receiving instruction using best practices.

    —Understanding their community and their own literacy.

    —Improvement of critical reading and writing skills.

A copy of the full proposal is attached.

Principal Investigator

_____

(Signature)

Date:_____

4

How Do I Find Answers? Carrying Out Your Qualitative Research Study, Part I— Reading, Observing, and Analyzing Artifacts

All of the planning you have done to this point will lead you to the heart of your research: finding answers to your research question. Starting the search for answers involves making strategic decisions about which methodological tools to use. In this and the next chapter, we identify several of these tools (reading, observation, interviews, surveys, and artifact analysis), explaining, in detail, how to use them and discussing their advantages and disadvantages. As you will discover, there are many reasons for selecting certain tools over others: from the nature of your research question to the applicability of a particular tool to a particular setting. Part of your journey in this chapter, then, is to determine which methodological tools will help you best answer your research question.

GENERAL ISSUES IN CARRYING OUT QUALITATIVE RESEARCH

The following general issues are important when carrying out qualitative research:

- Selecting the right tool

- Adopting the right attitude in carrying out your research
- Researching in multiple ways and with multiple methods

Selecting the Right Tool

Research tools are intended to be just what their name implies: instruments that aid you in obtaining information that will answer your research question. The tools you choose will depend on a number of factors. First, and most important, to consider is the research question itself. If your research question, for example, is focused on the reactions of particular people to specific situations, then you may select tools that will help you learn about individual responses—tools such as interviews or surveys. If your question is concerned less with people and more with the ways in which significant pieces of writing are structured, then you may select a tool, such as artifact analysis, that helps you critically approach the language and set up of the writing. And if your question is situated in the everyday activities of writers at work, then you may want to focus on what you see and hear in the setting, perhaps relying heavily on observation as your primary research tool.

A second factor in selecting particular research tools is the context for your research. As a researcher you will have certain kinds of access to that context, including the people in it. For example, if you are a teacher and your research setting is your own classroom, then you may have unlimited access to students, their written work, and their ideas. It may be easy to carry out observations, conduct surveys, interview students, and analyze their written documents. However, other contexts may afford you less access. For example, if you are researching a workplace environment, then you may have permission to attend some meetings and not others, and permission to talk to only certain workers. In this sort of context, you would need to take these constraints into account when selecting your tools. Thus, both the context for your research and your access to that context will be important factors.

A third consideration in choosing your research tools is your own level of comfort with and knowledge about the tools. This comfort and knowledge depend in part on your past experiences as a researcher and on your personality. For example, some of our graduate students find that they are very comfortable with interviewing because they have experience working in journalism or in public relations, which gives them confidence when they carry out an interview. Other students we work with have backgrounds in literature as undergraduate English majors or teachers of literature, so for them analyzing artifacts is a comfortable and familiar way to gather information.

Knowledge and comfort level, then, are certainly factors in selecting particular tools. However, researchers sometimes have to work outside of their comfort zones to discover the information that will best help them answer their research questions.

Learning about the various research tools and practicing those tools in low-risk set-tings—the kind of exercises you will do in this and the next chapter—can help you gain confidence with the tools. Carrying out an interview, transcribing the interview, and reflecting on how the interview went can help you to recognize your strengths and identify any challenges you might face in using that tool; we know from past ex-perience that this kind of practice can help you feel better prepared to use that tool in your formal research. So, whereas we encourage you to think about your past experi-ences as one way to select appropriate tools, we also encourage you to consider try-ing unfamiliar tools if they seem appropriate for your research.

There is one additional factor to consider when you begin selecting tools: your own theoretical stance toward knowledge. As stressed in previous chapters, it is vital for all researchers to think about what they believe in regard to how knowledge is created, transmitted, and uncovered. Because research is all about understanding information outside of our everyday experiences, our beliefs about the nature of knowing truly lie at the center of how we understand the information we are gathering. For example, as feminists and social constructionists, both of us view meaning as something that is constructed by participants in relation to a particular event or setting. Thus, for us, re-search is generative and dialogic, and we view our participants as key contributors to it. Selecting tools that honor this stance on the world is vital. For example, interviews are an integral part of research for both of us. And the way we conceive interviews is likely different from how other researchers conceive them who don't bring to bear these same theoretical stances. We see interviews as less a formal question–answer exchange (in which our questions set the tone and, in fact, the knowledge base for what will come) and more as a dialogue, an interaction in which the direction of the interview is mutu-ally constructed with our participants. Similarly, your own stances toward knowledge will certainly impact your research and the tools you select for it. In addition, adher-ence to particular research methodologies—such as ethnography, case study, teacher or action research, or autoethnography—will also influence the tools you select and how you use them (please see the sidebar to this section).

These four factors—the nature of your research question, the context of your re-search and your access to it, your experience and comfort zone, and your theoreti-cal stances—will be vital considerations as you begin to select tools. What's also important to realize is that there are no absolute "right" or "wrong" tools for re-search. And it may help to know this: Even experienced researchers grapple both with the selection of tools and with how best to use them. We learn continually as researchers: We grow in our comfort, we occasionally experience discomfort (even with what's usually comfortable to us), and we learn the limitations and strengths of certain tools and of our approaches to using them. Ultimately, it is less important to be wedded to a particular tool and more important to keep an open mind and to be sensitive to which tools will work best for your particular study. The prompt included with this section focuses on the selection of research tools.

Sidebar 1: *Kinds of Qualitative Research*

We want to briefly define some of the common kinds of studies carried out by qualitative researchers—in particular, ethnographies, case studies, teacher research, action research, and autoethnographies. These definitions are limited—you can easily find articles and even books that address each of these approaches—however, we wanted to provide you with at least some indication of what each entails.

Ethnography: Ethnographies are usually concerned with understanding culture and individuals' everyday experiences within a culture. Ethnographers immerse themselves in a site, generally spending considerable time there.

Case Study: Case studies are usually concerned with a specific event, activity, or individual within a setting. Case study researchers try to understand that event, activity, or individual in great depth, much like ethnographers, but are more interested in the particularities of their single case, rather than the whole culture.

Teacher Research: Teacher research is the systematic study of a teacher's own classroom, based in a question of concern that arises from that setting.

Action Research: Action research suggests a research study that not only seeks answers to particular questions that arise in a setting, but also leads to action steps aimed at creating change in that setting.

Autoethnography: Researchers who do autoethnography study their own backgrounds, personal histories, and present experiences as members of a particular culture as a way to make better sense of that culture.

Adopting the Right Attitude in Carrying Out Your Research

As you select particular research tools and begin your research, you will also want to adopt an appropriate disposition toward your research. Both we and our students have found that research is easier and more satisfying when we keep our sights on our purposes for the research—that is, on what we hope to learn. Research, as we have said before, is about searching and inquiring; it's an exploration. It is important to remember that, and it is also important to give yourself the freedom to search and explore. You may end up missing out if you get too bogged down in trying to

Prompt 1: *Developing a Research Plan.*

Consider the following scenario and write about how you would research it. What tools would you use? What resources would you consult? Next, share your thoughts with two or three of your classmates. Did you all come up with the same approaches? What are the advantages and disadvantages of the different approaches each member of your group came up with? What is each person's rationale for his or her approach?

> *Scenario:* You and three of your friends have decided to use an upcoming 3-day weekend to go camping. You don't know yet where you should go, but you're all very conscious of your time constraints (you don't want to spend most of the weekend on the road), and you all have similar interests: you like hiking and canoeing.

find "the one right answer." You might also overlook important findings by not being open-minded or flexible enough to notice them.

It may seem trivial, but the disposition you have toward both your research and your participants—how and what you think about each of them—really can make a difference. Are you inquisitive and curious, playful and flexible, patient and open-minded? Do you think of yourself as a learner? The sidebar for this section contains comments from one of our own students that suggest the role such dispositions can play in our research.

When you are in the midst of a research project, hopefully you will also find yourself thinking like a researcher—for example, always seeing connections between your research and the events occurring around you in everyday life, becoming so immersed in your study that you continually ask questions that relate to it, making a habit of jotting down your questions and the connections you see. Also, do not underestimate the role of serendipity in research. For example, when Cathy carried out her research on community organizing strategies, she began noticing numerous newspaper articles about teachers and schools that had achieved some kind of public change. At the time, she also became involved in a neighborhood effort that was started to prevent a large development project. As she participated in this effort, she kept a journal about her experiences, drawing connections to the literature she was reading for her research and the interviews she was conducting. One of our former students, Diane Benton, had similar experiences. Soon after she started her project on developing a more diverse curriculum for composition, she noticed a flyer for a conference at a nearby university that addressed this very issue. She also learned that her advisor knew two of the featured speakers for the conference, so she was able to meet and talk privately with them. They offered her advice and shared copies of their papers with her. This experience helped Diane get a very strong start on her project.

We would argue that taking advantage of the opportunities that naturally present themselves to you will make you a better researcher. You will learn to be open to new ideas, to reassess continually what you are noticing, and to always ask new questions. All of these things will lend momentum to your research and also contribute to its effectiveness.

Sidebar 2: *Student Testimonial on the Importance of the Right Disposition in Research*

The following is a testimonial from one of our students on the importance of having the right disposition in research:

Rhonda Copeland: Due in large part to my conservative, religious upbringing, I have always had a stubborn ideological bent. I knew what I believed and I was right in my beliefs. Although I would listen to others' ideas and allow them to have their own beliefs, if their ideas did not concur with my own, they were wrong. And if they chose to argue with me about those beliefs, I let them know they were wrong.

I am still stubborn and I still hold strong beliefs, but I now recognize in myself a willingness to entertain and explore new ideas. ... One semester I found myself in altogether unfamiliar territory. One of my classes, *Language and Politics*, covered issues that I had never considered. I found myself soaking in the beliefs and ideas of my classmates and of my instructor, accepting some ideas without question while rejecting others outright. I found myself thoroughly enjoying my immersion into new territory. But when it came to exploring this territory on my own, I reached for something familiar to ground myself in.

Since my second major as an undergraduate was history, when it came time to choose my own project for the course, I chose to research the history of the language rights movement in the United States. ... I was exploring new territory, but right away I made a mistake. I convinced myself that I knew what I was going to find; I had a preconceived conclusion in my mind. I decided early on in my research that there was no such thing as the language rights movement. My stubborn ideological bent got in the way of my exploration, and I ended up standing on a literary soap-box spouting out what the language rights movement would have been if it had existed. My instructor's comments had to shove the soap-box out from underneath me before I realized what I was doing. But that shove taught me a valuable lesson, one I carry with me into new territory: I always need to be cautious not to let my ideological bent get in the way without being justified. I know that when I explore new territory, I need to do so open-mindedly. With new challenges come new ideas, and I know now that I am ready to accept those ideas.

Researching in Multiple Ways and With Multiple Methods

Something else to consider as you begin thinking about the research tools to use is how to vary and combine those tools. In other words, you will probably end up using more than one tool, and, in fact, you will need to use more than one to lend credibility to your findings through triangulation. When you triangulate your findings, you examine the same event, situation, or data in multiple ways, and you do so to determine if each way reveals the same or similar information—to see, in other words, if the multiple ways of looking at an event or situation confirm what you are seeing. If they don't, which sometimes happens, you need to consider what might be causing the discrepancy (e.g., whether your own subjectivities and biases are coming into play somehow, or whether one of the approaches you are taking in your research may be causing distortion). This is discussed further in chapter 6, so here we simply emphasize that you should and will want to use multiple methodological tools in your research, both as a way of checking yourself and as a way of insuring the credibility of your findings.

Prompt 2: *Researching in Multiple Ways.*

Find and read an article in your field that reports a qualitative study. Pay particular attention to the methodology section of the article and look specifically at how the researcher(s) used triangulation. Did the author(s) use it and, if so, how? Was it addressed explicitly in the article? Also, in your opinion, what did the use of triangulation contribute to the research? What would have been problematic in the research if the researcher(s) had not triangulated the data? What else might the researcher(s) have done in this respect and why?

METHODOLOGICAL TOOLS FOR RESEARCHING YOUR QUESTION

This section presents the common methodological tools for qualitative research. (We address reading, observation, and artifact analysis in this chapter, and we address interviewing and surveys/questionnaires in the next chapter.) Our presentations explain what each tool entails and then address the purposes the tool can serve. We also offer practical advice for using the tool.[1] Finally, rather than save

---

[1]What we provide here is really just an introduction to the tools that someone who is doing qualitative research is likely to consider using. Our coverage of these tools is not comprehensive; rather, our goal is to provide enough of an introduction to and overview of them to help you make sound choices about which ones you would like to use in your research and how you might use them. There are numerous sources that cover each of these tools in much greater depth. Most tools even have entire books devoted to them (see the sidebar in this chapter for some examples).

these discussions for the ends of the chapters, as we do in other places in the text, for each tool we discuss how particular lenses or biases can affect how you use it. We also explain what's involved ethically in using each tool. Because you no doubt will combine tools, we encourage you, as you read these sections, to think about how you might use these tools in relation to each other. We also encourage you to review again, especially as you read through this and the next chapter, the methods sections from the proposals we included in appendix B of chapter 3.

The following are the methodological tools we discuss in this and the next chapter:

- Reading
- Observation and participant observation
- Artifact analysis
- Interviews (chap. 5)
- Surveys and questionnaires (chap. 5)

Reading

We begin our discussion of qualitative research tools once again with reading in order to reinforce what a central role it will likely play for you throughout the research process. Many researchers use reading as one of their primary methodological tools (the methods sections in the proposals in the appendix of chap. 3 offer good examples). What is different about reading as a methodological tool from reading as a tool to develop your research question or to plan your research is that it is usually geared at this stage more toward answering your research question than it is toward helping you define or narrow your question. In other words, it's about more than just developing a sense of what's already out there. You read at this point in your research in order to uncover answers to your question or to learn more about different aspects of it. You may also decide to read works from other fields to broaden your perspectives. Your reading thus becomes a central part of your inquiry.

For example, when Cathy sought to answer a research question about how teachers could better help the public understand literacy practices, she read in a number of related fields to find answers. By reading articles, oral histories, conference presentations, and books by environmentalists, public health advocates, and grass roots organizers, she was able to learn how professionals in diverse fields use community organizing strategies to change the public's understanding of certain issues—information that became an important part of her study. Reading, then, gave her a set of answers with greater geographic and historical scope than she could have gotten from only doing interviews and observations.

Sidebar 3: *Sources Addressing the Various Research Tools*

Observation and Field Notes:

Adler, P. A., & Adler, P. (1994). Observational techniques. In N. K. Denzin & Y. S. Lincoln (Eds.), *Handbook of qualitative research* (pp. 377–392). Thousand Oaks, CA: Sage Publications.

Emerson, R. M., Fretz, R. I., & Shaw, L. L. (1995). *Writing ethnographic field notes*. Chicago: University of Chicago Press.

Fontana, A., & Frey, J. H. (2005). Recontextualizing observation: Ethnography, pedagogy, and the prospects for a progressive political agenda. In N. K. Denzin & Y. S. Lincoln (Eds.), *Handbook of qualitative research* (pp. 729–746). Thousand Oaks, CA: Sage Publications.

Frank, C. (1999). *Ethnographic eyes: A teacher's guide to classroom observation*. Portsmouth, NH: Heinemann.

MacLean, M., & Mohr, M. (1999). *Teacher researchers at work*. Berkeley, CA: National Writing Project.

Power, B. (1996). *Taking note*. Portland, ME: Stenhouse.

Interviews:

Finley, A. (2005). The interview: From neutral stance to political involvement. In N. K. Denzin & Y. S. Lincoln (Eds.), *Handbook of qualitative research* (pp. 695–728). Thousand Oaks, CA: Sage Publications.

Fontana, A., & Frey, J. H. (1994). Interviewing: The art of science. In N. K. Denzin & Y. S. Lincoln (Eds.), *Handbook of qualitative research* (pp. 361–376). Thousand Oaks, CA: Sage Publications.

Perakyla, A. (2005). Focus groups: Strategic articulations of pedagogy, politics, and inquiry. In N. K. Denzin & Y. S. Lincoln (Eds.), *Handbook of qualitative research* (pp. 887–911). Thousand Oaks, CA: Sage Publications.

Seidman, I. (1998). *Interviewing as qualitative research: A guide for researchers in education and the social sciences* (2nd ed.). New York: Teachers College Press.

Wise, R. (1995). *Learning from strangers: The art and method of qualitative interview studies*. New York: The Free Press.

Artifact Analysis:

Bloome, D., Shuart-Faris, N., Power Carter, S., Morton Christian, B., & Otto, S. (2005). *Discourse analysis and the study of classroom language and literacy events: A microethnographic perspective*. Mahwah, NJ: Lawrence Erlbaum Associates.

Brown, G., & Yule, G. (1983). *Discourse analysis*. London: Cambridge University Press.

Florio-Ruane, S., & Morrell, E. (2004). Discourse analysis: Conversation. In N. Duke & M. Mallette (Eds.), *Literacy research methodologies* (pp. 46–61). New York: Guilford.

Goldman, S., & Wiley, J. (2004). Discourse analysis: Written text. In N. Duke & M. Mallette (Eds.), *Literacy research methodologies* (pp. 62–91). New York: Guilford.

Halliday, M. A. K., & Hassan, R. (1976). *Cohesion in English*. London: Longman.

Hodder, I. (1994). The interpretation of documents and material culture. In N. K. Denzin & Y. S. Lincoln (Eds.), *Handbook of qualitative research* (pp. 393–402). Thousand Oaks, CA: Sage Publications.

Huckin, T. N. (1992). Context-sensitive text analysis. In G. Kirsch & P. A. Sullivan (Eds.), *Methods and methodology in composition research* (pp. 84–104). Carbondale: Southern Illinois University Press.

Scheurich, J. J., & McKenzie, K. B. (2005). Analyzing talk and text. In N. K. Denzin & Y. S. Lincoln (Eds.), *Handbook of qualitative research* (pp. 869–886). Thousand Oaks, CA: Sage Publications.

Some Tips on How to Read for Your Research

Researchers should have some way of recording what they find in their reading, that is, some system essentially for keeping track of what they read and for drawing connections between the different ideas they encounter. Most people devise some personal way of taking and organizing notes from their reading. Some record their notes in reading journals or logs or in files on their computers, which allow for easy searching. We also know people who read with highlighters, some who write notes in the margins of texts, and others who refuse to mark a text at all because of how they view printed texts. Others devise elaborate systems for flagging pages in texts using different colored Post-it® notes.

Prompt 3: Sharing Note Taking Strategies.

Share with your classmates the strategies you use for taking notes when you read. Note the variations.

You should develop a method for keeping track of what you read that's convenient and flexible and works well for you. Whatever method you develop, we encourage you to keep complete bibliographic information (e.g., use RefWorks, EndNote, or some other software or system for documenting what you read). You will need this information at certain points in your research and will want it to be accessible. What's immediate and memorable now will be much less so when you begin another task or start reading other sources.

The Role of Lenses and Ethics in Reading

We started addressing the issue of how lenses influence what and how we read in chapter 3. We all have personal and theoretical lenses that influence what we choose to consider in our research and how we view those choices. This is also true of what we read. Our lenses influence what we select to read and what we take from our reading; however, if we keep an open mind, the reading we do can also influence and broaden our lenses. What matters is that we be willing to entertain new perspectives, even those with which we disagree. We need to be aware of our lenses and of how strong a hold they might have on us. What might our lenses prevent us from seeing? If we shift lenses, what will happen? How differently might we view things? These are all questions we should ask frequently as we read for our research.

Ethics also come into play with reading, particularly in relation to how we acknowledge what we read. Most obviously, we need to credit our sources by citing them properly. We need to be careful as well about how we represent sources, and mindful that we do not take what an author says out of context just to support our own viewpoints. Our sources are each contributors to a larger conversation and, as we would in any conversation, we need to treat these contributors with respect. That doesn't mean that we can't disagree or take issue with an author, but it does mean that we should always strive to be fair and accurate in how we present and discuss an author's work.

Observation and Participant Observation

What Observation and Participant Observation Are

One of the most essential skills you can have as a qualitative researcher is the ability to observe situations carefully and to derive meaning from your observations. Almost all qualitative research involves observation, whether it's just informal observation of a setting to determine what to focus on in your research or more formal and systematic observation to collect data for your study. An important distinction to be made about this tool for gathering data has to do with your role as an observer. For example, you might be present in the setting just to observe and thus lack any formal or official connections to it, or you might somehow be involved in the setting as an insider or participant. If you are the latter, what is known as a participant observer (e.g., an employee, consultant, advisor, teacher), it is likely you will have a very different perspective on the setting than if you are there as an outside observer.

What is vital for any kind of observation—whether or not you are a participant in the setting—is the relationship you establish between the familiar and the unfamiliar. Sometimes, especially if you are a participant observer, your job will be to make the familiar unfamiliar—to take the everyday events that are so comfortable and recognizable to you and to learn to see them with new eyes. At other times, more often when you're an outside observer, your job is just the opposite—to be able to take a new setting and all that is unfamiliar and to get to know it well enough so that it becomes familiar. Both vantage points, as we will discuss in detail shortly, have advantages and challenges. For example, as a participant you may have access to settings you otherwise would not have access to and the ability to see and experience those settings in a manner similar to an insider. However, your insider status may prevent you from noticing important aspects of the setting. It is necessary, therefore, to understand your positioning and to be aware of what it does and does not afford you as a researcher.

When and Why You Observe

There are any number of reasons why you might observe a setting. For example, many researchers carry out observations early in the research process as a way of formulating a research question or as a way of familiarizing themselves with a setting to see what might be interesting in it. Once a study is underway, observation becomes a useful tool for obtaining a wide variety of information. It can reveal:

- How people interact—what the social dynamic of the setting is.
- What people do in the setting—how they approach and perform various tasks.
- How they respond or react to various experiences.
- What occurs regularly or only sporadically in the setting.
- How the setting is structured and organized.
- When and for what reasons the structure or nature of the setting changes.

When carried out repeatedly, observations can provide a researcher with a first-hand view and record of what occurs in a setting over time.

How You Observe

Observation, for a number of reasons, is often trickier than it seems. One challenge has to do with focus. Faced with a complex setting that has a lot going on (a classroom filled with 30 teenagers, some of whom are focused on the teacher, some of whom are passing notes, some of whom have their heads down on the desk; or a meeting of 75 employees, some of whom are listening intently to the speaker, some of whom are surreptitiously reading or doing other work), how can you ever know what to write down in your observation log? You obviously can't see everything at the same time, so how do you decide what to observe? And then what happens if you focus too intently on one aspect of the setting? Will you miss other events that might be important to your research?

Another challenge has to do with your role as observer, as mentioned earlier. If you are a participant in the setting, how will you recognize as important those things you might take for granted because you are so familiar with the setting? As a participant observer, in particular, you face the challenge of being part of a culture while maintaining enough distance from the culture to make sense of it. As we noted previously, your involvement can be both an asset and a liability: It may allow you to understand the setting in a way that an outsider never could, but it also can prevent you from seeing the most interesting aspects of the setting. Lucy Calkins (1983), a teacher researcher, offers a particularly telling anecdote that illustrates this challenge. In talking about her first-ever experience as an observer in a classroom, she says:

> On my second day as a researcher, Don Graves joined me in Pat Howard's third-grade classroom. The children weren't writing, but Graves suggested we stay. I paced up and down the rows. The kids were all copying things out of their math books. I anxiously waited for someone to *do something* so I could gather some data. But no, they just kept on copying out of those math books. I went to the back of the room and leaned against the radiator to wait for some data to appear. Nothing. Finally I signaled to Graves, who'd been scurrying about, and we left.
>
> Before I could let out a quiet groan, Graves burst out with, "What a gold mine! Wasn't it amazing? How'd you suppose that one kid up front could write with a two-inch pencil? And that guy with the golf ball eraser on the end of his pen. Zowie." In his enthusiasm, Graves didn't notice my silence. …
>
> I had learned a big lesson. The task of case-study research is to make the familiar unfamiliar. (pp. 9–10)

If you have ever been a participant observer, or even an outside observer, you may be familiar with Calkins's experience. It is difficult to observe a group of people in a setting and recognize anything unique in it, anything that seems worthy enough to call data. As an observer, however, your main task is to see what's in front of you with new eyes and to try to understand what you see from the perspectives of the participants. Clifford Geertz (1983), an anthropologist who has written extensively about observation, suggests that the best way to understand what's really going on in a situation is through "thick description." By this he means detailed description that gets beneath the surface and that allows you to get at the meanings behind everyday events. Geertz discusses the importance of situating your observations in the "native's point of view," of trying to grasp the meaning of the situation by "putting oneself into someone else's skin" (pp. 57–58). He says, "The trick is not to get yourself into some inner correspondence of spirit with your informants. … The trick is to figure out what the devil they think they are up to" (p. 58). As Geertz and many other researchers contend, such understanding comes only over time with multiple exposures to a setting.

Prompt 4: *Carrying Out an Observation.*

If you are reading this text as part of a class, select a site (a mall, a lobby, a waiting room, a restaurant) and go to it with a couple of your classmates (together or separately). Take a notebook and pen and record your observations for 30 minutes (each of you should do this separately). Try to do what Geertz suggests and develop a "thick description" of the setting. Once you have all done this, share your descriptions and talk about what all of you noted and what only a few of you noted. Also talk about why you chose the things you did to record and about what you did to get at the meaning behind the events and to get at the points of view of the people you observed.

Probably the most important aspect of doing observations is keeping field notes, which are the notes you write down either during or immediately after an observation. Most researchers take at least some notes as they are observing and then fill in those notes as soon as possible after the observation. Some, however, wait until later to record their observations, feeling that doing so during an observation makes participants self-conscious or interferes with the observation. We recommend that you do some of each, mainly because of the challenges we see with relying solely on writing notes after-the-fact. First, researchers often discover that their memories of what they observe are not reliable. The time gap can lead to forgetting important moments, reporting a verbal interaction incorrectly, or even creating details that didn't occur in quite the way they were recorded. Second, the time gap can also change your note taking from simple recording to recalling and recreating.

Whatever approach you decide on, you should know its strengths and its limitations. Sometimes the situation won't allow you to take notes as it's happening and the after-the-fact notes will be the best you can do. And there are other challenges you may face in recording your observations. For example, if you are a teacher observing your own classroom, how can you teach, observe your students, and take notes all at the same time? Or, if you are a writer in an organization, how can you carry out your job functions, observe the setting, and also record your observations?

Another challenge you may face is determining what to record. Should you record every conversation, note every movement your participants make, keep track of every event that occurs while you're observing? Doing all of these things can be overwhelming. On the other hand, should you only record those moments that connect directly to your research question? This approach also has its limitations, especially if your research question eventually changes and you find that you have not recorded information that is now vital to your research.

Yet another challenge speaks to a key concern in qualitative research: How trustworthy are your observations? As impressions you record as a single observer, can you trust their accuracy? Can you trust that they reflect the events as they occurred? Despite your attempts to understand the site from "the native's point of view," as Geertz suggested, what you record is ultimately your personal take on the setting. How can you be certain that you are really capturing what's happening there?

Although we urge you to keep these considerations in mind, our best advice is to just jump in and observe. The following are some of the more common types of information to record from your observations:

   • *The physical set up of your research site.* Questions you might respond to include some of the following: What does the space look like? If it's a classroom, meeting room, or office workspace, how are the chairs, desks, and tables arranged? Are there computers, video screens, other pieces of equipment? Are there windows, works of art on the walls, a clock? What's the temperature in the room? Does it seem too warm or too cold? How many people are in the room? Is it crowded? Is it spacious? What does it sound like? Although some of these questions may seem peripheral, being able to recall the physical makeup of the site can help you with your final write-up: You can make your write-up more persuasive by offering your reader specific narrative details along with a sense of "being there."

• *The interactions that occur among participants.* Some questions you might respond to include the following: What is actually going on with and between people in the room? Are they talking? Are they writing? Are they staring at the clock? Are they whispering? Is one person in charge? What is the person doing? If conversation is ongoing, who's talking? Who's listening? What kinds of body language are being exhibited? Observing how people interact and behave provides rich material for your research. If you are studying the revision habits of student writers, for example, watching what the writers actually do during an in-class revising session can give you vital information.

• *Snippets of the conversations that occur.* Some questions here include the following: What are people saying in the room? Are individuals responding to questions? Are people talking quietly among themselves? Are they taking turns in the conversation and talking more formally? Although you can't possibly write down verbatim everything that everyone says in a situation , you may be able to capture a few moments of conversation that seem interesting: a boss's charge, a student's response, a worker's complaint, an outsider's interruption. Although this kind of information is not the same as a transcription of an interview, a general sense of a conversation can help you capture the feelings of the moment and perhaps provide you with a jumping off point for a later interview.

Done well, your observations can be an incredible resource. The information you obtain from them, as the previous list suggests, may speak directly or indirectly to the question that is guiding your inquiry. Observations also provide important narrative detail and background information.

A final consideration with observations concerns where to record them. Observation notebooks provide a central place for recording the reactions and understandings you gain through your observations. If you keep a notebook, then you should develop a routine for writing in it and follow that routine faithfully. Learning about the many kinds of research notebooks you can use to record your observations (see the sidebar for this section) can help you make an informed decision both about what you record and about how you record it.

Prompt 5: *Selecting and Observing a Setting and Recording Your Observations.*

Brainstorm a list of potential observation sites for your research. What advantages might each of these sites offer your research? What challenges might they pose to you as a researcher? Select one site and carry out a preliminary observation. Try one of the note taking strategies addressed in the sidebar to record your observations. Report back on your observations and on your experiences with the type of note taking you tried.

Sidebar 4: *Different Ways to Record Observations*

There are many different ways to record observations that you might find useful. The one you pick will depend on a number of factors: your research question, your role as a researcher, and your level of comfort. We encourage you to try some of the options we present here and adapt them as necessary. All of these, for example, can be done electronically, especially if you have a laptop that you can take to your research site. Experiment to determine what kind of tool or notebook, whether paper or electronic, will work best for you.

• *Traditional Observation Notebook.* This is a notebook in which you write down your observations as they are occurring. You date the entry at the top and keep running notes (often with times noted in the margins) throughout the observation session. These logs tend to be reportorial in nature. If you use this type of log, you may add a chart showing the physical layout of the space you observed or a tally sheet that records recurring events (e.g., how many times each participant spoke).

• *Two-Notebook Approach.* This approach entails keeping two notebooks, one for recording observations and one for recording your responses to what you have observed. As an example, Cathy adopted this approach some years ago. After keeping a running log in a classroom observation setting for months, she discovered that the log did not allow her to summarize, react to, or even vent about some of the practices she was observing. Even though she knew that all observations reflect our own biases, she tried to confine the running log to what she noticed, not how she felt about what she noticed. She began to keep a second notebook that she never brought to the research site where someone might ask to read it. In her at-home notebook, which she wrote in once a week, she summarized what she was noticing so far and laid out how she was feeling about her observations, the human reactions she was having. This second notebook became a place where she summarized what she observed, raised new questions, and reflected on how what she was observing related to those new questions.

• *Dialogic Notebook.* In this notebook, you write running notes of your observations on one side of the page and comments, summaries, and responses on the other side. This system lets you generate both the detached notes, characteristic of the running log, and the more personal responses you might record in a second notebook. The major advantage, of course, is the ease of having just one notebook. Moving back and forth across the pages is much easier than keeping track of two notebooks. The downside is the public nature of such notebooks. If you are observing in a site in which the participants are a bit wary of what you are recording ("What are you saying about me?"), then you might want to have a log that you don't mind sharing with them. If you use a dialogic notebook, then you need to be careful about what you write in your comments.

• *"In the Middle" Notes*. These are designed for situations in which researchers do not have time to keep a running log. With this approach, you might take 3 to 5 minutes in the middle of your activity to jot down what you noticed so far that day. (If you are including your participants' voices as part of your study, then you might also ask them to write their response in their own notebooks.) You can also put your notes on Post-its® and place them in your notebook at the end of each day. (This option is adapted from the work of Ruth Hubbard and Brenda Power.)

• *"After the Fact" Notes*. These are comparable to "In the Middle" notes, except that you do them at the end of the observation. In these notes, you summarize what you remember. Although these have the disadvantage of being done after-the-fact (and thus have the potential to lose some of the immediacy of your observations), they still can be useful. (These, too, are adapted from Hubbard and Power.)

• *Tape-Recorded Log*. Finally, some researchers tape-record their notes at the end of a long day. They may do this during their evening commute or even in an easy chair at home. Although you lose some of the immediacy of the running log, you still can capture the highlights of the day before you forget them. Keep in mind that if you use a traditional tape recorder, you need to transcribe the tapes, which can be a time-consuming process. On the other hand, transcription can give you time to really process the observation mentally and to begin thinking about the salient aspects of it. You might also, if you have the right equipment and software, record your notes directly onto your computer. This can make listening to and reviewing them fairly easy, especially if you burn the file onto a CD or download it onto an iPod or other MP3 player. You will need, however, to transcribe or convert the voice files so that you have text files of your data.

The Role of Lenses in Observations

As with all of the tools addressed here, what you end up seeing with your observations will depend on your personal and theoretical lenses. Here is where the lens metaphor is perhaps most explicit: What you see and how you see it depend on the lenses through which you view it. In fact, you probably discovered in the observation prompt that you completed with your classmates that no two of you observed the same thing in the same manner, or reported it in the same way. This would be true of any observation, but it is especially true of the observations we do for our research. Again, what's important is being aware of your lenses and how they might influence your observations. It's good to have some sense of why you are seeing what you are seeing and how your personal and theoretical perspectives may be influencing your interpretations of what you see. It is also good to be aware that there may be things you fail to see as a result of your personal and theoretical lenses.

Sometimes, when possible, it can also be helpful to check your observations against those of another researcher or observer in the setting. This is known as member-checking. Some classroom teachers, for example, ask a student volunteer to take observation notes on a particular day. Similarly, in a workplace setting, a researcher may ask an employee to take such notes. Checking your observations against those of the student or employee can reveal how your personal lenses might lead you to see the same event differently from how others in the setting might see it.

Observing Ethically

Observations also have numerous ethical dimensions. You need to think, first, of who you are observing. Are they adults who are fully aware that you are observing them? Are they children in a classroom? Are they participants in an online discussion who are not aware that they are being observed? You also need to think about where you are observing these individuals: Is it in a public setting, a private one, a school, or an online environment? Do you have permission to observe in that setting? A third consideration is how you will use your observations and what, if anything, you will tell the individuals you observe about how you will be using your observations. How might these individuals be affected by your observations?

It is our belief that you should always be up front with the individuals you observe. You should inform them that you are observing them, you should let them know what you intend to do with your observations (how you will use them), and you should offer them the option of not being part of your observations. Some people are very self-conscious about being observed, even to the point of it influencing their behavior (what is sometimes called researcher effect). As one possible exception to our advice, if you are observing in a public setting—for example, if you are just observing generally how people respond to vendors in the concourse of a mall—you may not need permissions. However, if you are doing more than that—for example, recording and disseminating specific information about particular interactions—you definitely should obtain permissions.

Being ethical in the observations you do depends, at least to some extent, on your own reading of the situation. Again, you should think about what is fair and responsible in the situation. Certainly, you will need permissions if children are involved. In most cases, you will also want to seek permissions with adults. Further, if someone asks you not to use something that you observed, then you should respect the request.

A final ethical consideration has to do with the claims you can make from your observations. You ultimately may end up wondering, as we noted previously, how reliable your observations are, or to what extent you can trust their accuracy. Your personal and theoretical lenses ultimately will influence what you observe and how you interpret and make meaning of it. That's inevitable. What becomes problematic is when your lenses blind you to other interpretations that may also be valid. In the final analysis, you probably can never be sure that you are capturing everything that's happening in a situation; however, you can at least acknowledge your doubts, and you can acknowledge the role your lenses and biases may be playing in influencing your observations.

Artifact Analysis

What It Is and Why You Do It

In addition to carrying out observations, you may also end up collecting and analyzing artifacts that connect to your research question. The term *artifact* seems scientific, but it is an apt one for qualitative research. Anthropologists, for example, often collect tools and other material artifacts that shed light on the work or social behaviors of a particular culture. An artifact is essentially physical evidence that researchers examine to better understand the issues and people they are studying. Writing researchers doing historical research may collect artifacts like various writing tools, but most commonly we collect and analyze the written documents that the people we study have produced or worked with. For example, when Ann carried out her research with the physicists, she collected every draft they produced of two articles they were writing. This amounted to 23 drafts of one article and about 14 of the other. She also collected all of the drafts of a related conference paper they had written, and she obtained copies of some of their other publications and grant proposals. Her analysis of these documents constituted a major part of her research.

Usually, the artifacts we select to analyze are determined by our research question. For example, if you're trying to determine how individuals compose, like Ann was, you probably want to collect drafts of their papers as they are writing them. You might also be interested in particular aspects of the composing process, like planning or revision, and so focus your analysis on just the documents writers produce at those stages. Or, you may be interested in how writers compose in particular circumstances or settings, or how they write specific kinds of documents, like the scientific journal articles Ann studied.

Another common concern writing researchers have is with examining the traits and features of various kinds of documents. In these cases, researchers usually focus on texts that are already completed or published; however, their purposes for doing this may vary. For example, some researchers examine texts with a concern for the various rhetorical and linguistic strategies the authors used; others focus on argumentative strategies; and some examine the difficulties writers have (e.g., particular problems that appear regularly in certain writers' texts). You might also examine documents to infer things. For example, one of our students analyzed call logs from the customer service department of her company to determine the difficulties users were having with one of the company's software products. She wanted to determine how the online help for this software might be improved, and this was one of the only sources of information she had available to her (because the writers in her department lacked direct access to users, the call logs were the next best option).

Prompt 6: *Looking at How Others Analyze Artifacts.*

Find a research article in your area of concentration that relies on an analysis of documents (e.g., student texts, workplace documents, etc.). Read the article and prepare a report for your classmates in which you identify and discuss what documents the researchers analyzed, why they analyzed them, how they analyzed them, and what they discovered or learned from the analysis (what arguments or claims they make based on the analysis).

Artifact analysis can also be informative prior to your research. Many researchers, in fact, use it as a preliminary data-gathering strategy to get a sense of what people are writing, how they are writing it, and what kinds of writing are typical, or not, in a setting. So instead of obtaining information primarily from people, with artifact analysis you focus more on things (documents), although people certainly may be involved (e.g., in interviews focused on the artifacts), which is discussed further later.

How You Do It

Artifact analysis can yield a great deal of information, so usually researchers narrow it in some manner (e.g., they look at only certain features of texts, like the rhetorical appeals of ethos, logos, and pathos). How you decide to focus your analysis of artifacts depends on the theoretical tools you use for your analysis. There are numerous theoretical tools available for analyzing artifacts, some of which are addressed here.

Before deciding how to analyze the artifacts you collect, you need to determine which ones to collect. Some documents that writing researchers frequently analyze include the following:

- Student papers.
- Journals.
- In-class writings.
- Web sites.
- Instructional texts.
- Online help files.
- Magazine, newspaper, or journal articles.
- Meeting minutes.
- Reports.
- Proposals.
- Notes.
- Curricular documents.
- Call logs.
- Memos and letters.
- Books.

In qualitative research, you will usually obtain these documents directly from your participants. Sometimes, however, you might have to look through company or school files, search the Internet, visit a library, or even purchase the documents from a bookstore or publisher. Because it is not always possible to anticipate what documents you will need or find useful for your research, you might decide to collect more than you will need. You may also want to make copies of every document you collect so you have an unmarked back-up should you need it.

Prompt 7: *Selecting Documents to Analyze.*

Review the list of documents that writing researchers frequently analyze. Given your research question and the setting in which you are planning to do your research, what documents might you select to analyze? Share your list with a classmate or in a small group and brainstorm other possibilities.

Sometimes writing researchers shy away from artifact analysis because they are afraid they lack the skills they need to do it properly. Few of us, for example, have an extensive background in linguistics, yet some sort of discourse analysis is often called for in analyzing artifacts. Those of us trained recently in composition studies may also not have an extensive background in literary analysis. Most writing researchers, however, do have experience analyzing both the surface and deeper meanings and features of texts. In other words, the knowledge you bring as a student of composition and rhetoric should be sufficient for doing this sort of work. Further, you can read and analyze written artifacts in many different ways in qualitative research. You should select those ways that work best for your research and that you are best qualified to use.

We present here several common ways to analyze documents. All of these approaches have books and even entire courses devoted to them, so what is presented here is simply meant as an introduction and overview. Hopefully, this review will give you enough information to select and begin using the approach (or combination of approaches) that will work best for your research. We encourage you to consult other sources that address the approach you decide to use (see the sidebar that lists sources on research tools). Also, remember that each approach suggests and makes use of various theoretical perspectives, which are addressed in the lenses section.

The following is a sample text that we refer to throughout our discussion of these approaches. This text is a response to an in-class prompt where students were asked to write an in-class essay about the topic of stress. The writer was a high school sophomore:

> It appears to me that stress is rather good for a person. However unreasonable amounts of stress can cause some mental anguish. I write this because at this moment I cannot think of nothing stressful that has happened to me. I am only fifteen (15) years old and I have not yet experienced stress in it full force.
>
> As I think about all the problems that children my age are facing, such as: rape, divorce, child abuse, unwed mothers or fathers, poor, welfare, I see how much I am blessed by not having this stress upon me.
>
> But I do recall one stressful time in my life. I was in the fifth grade. It was winter time and my cat came home sick. He (his name was Cinderella) had a small bruise on the top of his head. I didn't thing much of it at the time and I went ahead and fixed his milk for him to eat. Cinderella stayed to the house that night and died, right out on our back porch. When I came down stairs the next morning my cat was just as hard as a brick and cold as ice. It took almost a week for mom to get me another cat. Not exactly stressful but it is a story.

Rhetorical Analysis

One of the most common types of analysis in which writing researchers engage is rhetorical analysis. Analyzing documents rhetorically involves considering the document within the larger context in which it was created and/or in which it functions. This context includes the author, the author's purpose in writing (the exigence or rhetorical situation for the document), the manner in which the document was written (e.g., collaboratively, with reviews by higher ups), the audience(s) for the document, the mode of delivery, and so on. In analyzing a document rhetorically, you will likely make use of a particular theoretical concept within rhetoric. For example, you might use stasis theory from classical rhetoric or the concept of kairos, which has to do with appropriateness and timing. You might also analyze documents for their rhetorical appeals, drawing on Aristotle's notions of ethos, pathos, and logos. Or, you might use concepts from more contemporary rhetorical theory (e.g., Burke's pentad, Bitzer's notion of rhetorical situation, Perlman's new rhetoric). One of our students, for his master's writing project, was interested in how metaphors are used in first-term presidential inauguration speeches. His proposed corpus included all of these speeches dating back to Abraham Lincoln. He planned to draw in his analyses on Lackoff and Johnson's and other scholars' work on metaphors, and he planned to count "novel or innovative metaphorical expressions." His purpose was to compare the presidents over time and to relate the metaphors each president used to their visions and goals for their presidencies.

In analyzing documents rhetorically, you might choose one of these particular theories or you might focus on certain rhetorical features of a document. You might consider, for example, why a document is composed the way it is, what it is intended to achieve and with whom, how it strives to achieve that, whether or not it is effective, and why it is or is not effective.

If you analyzed the piece of student writing presented previously from a rhetorical perspective, you might choose to focus on the rhetorical situation of the document; that is, the reasons the author had for writing the document. If you did, you probably would note the following. First, the student knows this is a piece of "school writing" and seems to have a good sense of what has to go into such a piece. She knows, for example, that a school piece is supposed to move back and forth between generalizations (" … stress is rather good for a person. However, unreasonable amounts of stress can cause some mental anguish") and specifics (the incident recounted in the third paragraph about her cat dying). You might also note, however, that her knowledge of what makes school writing successful is complicated by the fact that she is struggling to find something to say. Directed to tell a story about adolescent stress, she can't seem to find a way in. She knows what stressful stories are supposed to be about (e.g., rape, divorce, child abuse), but she also knows that none of these things has happened in her life ("I am blessed by not having this stress upon me"). Thus, she appears almost baffled about what to do with the assignment.

Ultimately, what the student tried to do here is what she knows is required of her. She recounts a story about the most stressful event she can remember in her life, even though she's aware that it really is not that stressful in the larger scheme of things. She proves in her recounting that she is also aware of some of the strategies for effective storytelling: narrative ordering, specific details ("my cat was just as hard as a brick and cold as ice"), and the importance of conclusions ("It took almost a week for mom to get me another cat"). She also makes it clear, however, that she is aware that this is not what her teacher has in mind for the assignment: "Not exactly stressful but it is a story."

Linguistic or Discourse Analysis

Another approach that writing researchers often take in analyzing documents is to do a linguistic or discourse analysis of them. This entails looking at language use. There are two ways to carry out such analyses. One is to consider either the text as a whole or parts of it alone as independent objects of study. The other, which is referred to as context-sensitive text analysis, entails taking into account the setting and situation in which the text was created or is functioning (see Huckin, 1992). In either case, you consider the linguistic or grammatical features of the text in order to discern patterns, errors, recurrences, or other interesting linguistic information.

There are now numerous computer programs that will automatically parse and analyze texts linguistically. One common program that does this is Monoconc (http://www.athel.com/mono.html), which the graduate student we mentioned earlier also planned to use. His goal for this aspect of his analysis was to determine what pronouns appear in the inauguration speeches and with what frequency (e.g., first person *I*, second person *you*, and third person *we*). He also wanted to examine each president's use of modal auxiliary verbs such as *must*, *will*, and *can*, and word frequencies. To do these analyses, he planned to perform this computerized corpus-based analysis. His research questions, excerpted from his research proposal, offer a sense of how he linked the linguistic and the rhetorical features of these texts:

> How do individual presidential inaugural speeches differ over time in terms of their reliance on metaphor? How does the usage of personal pronouns, modal auxiliary verbs, or other linguistic elements vary? In addition to individual differences among the presidents, what will comparisons between political parties reveal? … In short, what are the rhetorical and linguistic components of this type of political speech and how do they vary over time?

He began his research by doing extensive background reading in each of these areas (texts and context, metaphor, and personal pronouns).

Linguistic analysis is of interest to writing researchers for many reasons. For example, if you were interested in a writer's development over a period of time, say a semester, it might be useful to see if the number of complex sentences the writer uses increases as that semester progresses. In their study of spelling, Rebecca Sipe, Dawn Putnam, Jennifer Walsh, Tracey Rosewarne, and Karen Reed-Nordwall focused part of their analysis on the number of words the students spelled incorrectly, the kinds of spelling mistakes they made, and the changes in their spelling over time. All of this provided useful information for their overall study (Sipe, 2003).

Finally, linguistic or discourse analysis is often coupled with other research tools, like interviewing. Several years ago, Lee Odell and his colleagues described a method called discourse-based interviews, which has proven quite useful to writing researchers, especially those who study writing in the workplace (Odell, Goswami, & Herrington, 1983). With this method, which can easily be adapted, the researcher analyzes the text for various elements and then interviews the author to determine why the person chose to use (or not use) those elements. These elements might be common linguistic or rhetorical devices, certain words or punctuation, common expressions, and so forth. The interviewer might also ask the participants if they had thought to use a particular element and then why they chose not to use it. This approach captures authors' reasons for using or not using different language elements, and it sheds light on their decision-making process.

If you chose to do a linguistic analysis of the student text shown previously, you could approach the text in many ways. For example, you might wonder if there is a correlation between the kinds of sentence structures the student uses and her struggle to find ideas. Here you might investigate whether she relies more on simple sentences or on complex sentences as she searches for a central thesis in the first two paragraphs. You might also consider if there is a contrast in sentence structures in different parts of the text. A simple quantitative analysis would reveal that, overall, she wrote seven simple sentences, four compound sentences, and three complex sentences. Her first paragraph begins with two simple sentences and then moves to one complex and one compound sentence. Her second paragraph contains a single complex sentence with 39 words. In paragraph three, she has nine sentences—the most yet—but her sentence structure does not reveal anything startling: She uses five simple sentences, three compound sentences, and one complex sentence. Overall, such a simple quantitative analysis may be one piece in a larger examination of this student's writing over time, comparing, for example, her writing in this more formal testing situation to less formal writing.

In another kind of linguistic analysis, you could look at the register of various words and phrases in the student's text: Are they formal or informal; are they Standard English or dialect? With this kind of analysis, you might make note of two particular points in the text. In the first paragraph, the student indicates that she's "only fifteen (15) years old. ... " Her inclusion of the parenthetical may lead you to wonder whether she is trying to formalize the piece, making it "sound like" certain genres of writing with which she might be familiar. She might also have used this just because she is not sure which construction is correct and so wants to cover all of her bases. A second interesting construction is found in the third paragraph where she states, "Cinderella stayed to the house that night." "Stayed to the house" is an unusual expression in a piece that is mostly written in Standard English. Why might she have used this construction? One way to find out would be to interview the student to learn more about the reasons behind her various choices.

Thematic Analysis

A third way to analyze documents is to look for themes that connect to the question you are researching. For this kind of analysis, you would focus less on the language and grammatical constructions in the text and more on its content. Thematic analysis can help you see if a particular idea or subject recurs in the writing or if there are particular places where contradictions arise.

For example, in the research mentioned earlier that was conducted by Rebecca Sipe and her colleagues (2003), thematic analysis was a major research tool. Early in their study of middle and high school students' spelling, they invited students to write "spelling histories": short pieces in which the students wrote about early memories of spelling, how they learned to spell, and what attitudes they had noticed others holding about spelling. Whereas they also analyzed these histories for types of spelling errors (a type of linguistic analysis), their primary focus was on the themes and issues that the students raised in their writing and the commonalities that existed in the students' spelling experiences. Looking across the entire set of spelling histories, the researchers drew conclusions about common traits they found for what they termed "challenged spellers."

Applying thematic analysis to the student paper included in this section could provide you with some insights into her thoughts about adolescence and stress. The text seems to indicate, for example, that she believes real stress does not occur for most 15-year-olds. The text also shows that she equates stress with extremely traumatic life experiences like "rape, divorce, child abuse, unwed mothers or fathers, poor, welfare." These are experiences she sees as causing real mental anguish and are those that she says are far removed from her own life and from her own experience with trauma ("my cat was just as hard as a brick and cold as ice"). Expanding this look at a single student's paper to the responses of an entire class to the same writing prompt could lead to further understanding of what adolescents consider stressful in their lives.

Whatever approach you choose to analyze documents, remember that you should not rely too much on just a single text to give you answers. Most researchers collect a number of related artifacts (a set of student papers, a portfolio of one student's work over the course of a semester, several drafts of an article or paper, several related Web sites, call logs from at least a few months, multiple articles addressing a single topic) and then analyze them as a group, searching for similarities and differences across the texts. Analysis, especially rhetorical and thematic analysis, then becomes a process of reading and rereading the texts, looking for features and statements that help you better understand your research question.

Prompt 8: **Analyzing a Document.**

Select a document connected to your research question. Try out each of the three approaches to artifact analysis talked about in this chapter: rhetorical, linguistic, and thematic. What insights does each approach give you? What questions does it raise? What further information might you need or want?

The Role of Lenses in Artifact Analysis

Lenses also play a central role in artifact analysis. The personal and theoretical lenses you hold will determine both what artifacts you choose to collect and how you choose to analyze them. These lenses will ultimately also influence what you find in your analyses and what meaning you make of what you find. Artifacts can suggest many different things, so you need some way of focusing your analysis, which the various methods discussed in this section will enable you to do. Those methods foreground certain aspects of the artifacts, while subordinating others. And, presumably, those aspects you foreground will connect in some significant way to what you are investigating.

Of course, as with any of the tools already discussed, you need to be aware of the lenses you hold. Because your lenses will lead you to view the artifacts you examine in certain ways, they may prevent you from noticing certain other things that may be pertinent to your research. In some cases, you can guard against this by having another person use your analytic approach with one or a few of the artifacts you have collected. (This approach is especially useful when you are using some of the more subjective analytic approaches, like rhetorical and thematic analysis and certain of the noncomputerized linguistic analyses.) You can then see if they found the same things you found. If there is a significant discrepancy, then you will want to consider why the discrepancy occurred.

It is also important to remember that your lenses may change during your research. With artifact analysis, what you notice when you analyze a document at one point in your study might be quite different from what you notice later. In other words, you may eventually look at the same artifact and notice different things about it. Presumably you are learning and gaining knowledge as you move forward in your research, so how you look at things, and what you end up seeing, will very likely change as you make progress with it. Everything about the research process is dynamic, including, and perhaps especially, the personal and theoretical lenses you hold.

Analyzing Artifacts/Discourse Ethically

What is probably most important to remember as a writing researcher is your obligation to the individuals who composed the documents you analyze, especially when those individuals are students. For one, you should never analyze a student's work and then report your analysis publicly without the student's permission (and/or their parent's permission if the student is a minor). Students have a right to know, first, that you are using their work for research purposes and, second, how you are using it or what you plan to do with it and why. This is called full disclosure. Even if you are analyzing documents that you found on the Internet, or e-mail messages that individuals posted to a listserv, then you should seek the permission of the authors of those documents. Make sure it is okay with them that you are using their words in your research and let them know how and why you are using them. Most individuals will have no problem granting permission for this sort of thing, and certainly it is better to have their permission than to hear back from someone whose work you didn't obtain permission to use. Also, in all of these cases, you should make the authors of the texts anonymous, assigning them pseudonyms so that they cannot be identified. Additionally, remember that when you analyze a written document, you are usually not doing so to judge it as being "good" or "bad." Your purpose in the majority of cases is to describe and analyze it, not to critique it.

With already published texts, the ethical issues are slightly different. For example, scholars in the rhetoric of science commonly examine already published scientific texts, like Watson and Crick's work announcing the double helix structure of DNA. With this kind of work, scholars need to be sure to properly cite the original texts and, also, if they reproduce significant parts of the original texts in their publications, they need to obtain the appropriate permissions from the publishers.

Finally, ethics in artifact analysis also concerns what you choose, or do not choose, to analyze, how you decide to analyze it, and what significance you attach to your analysis. For example, you may choose certain documents and not others, and in doing so exhibit certain biases. You may also need to be concerned with how many documents you analyze. In some cases (e.g., when you are doing computerized linguistic analysis), the size of your corpus will be important for the credibility of your findings. If your corpus is not large enough, then your findings won't be reliable. In other situations, a smaller corpus may be sufficient. Again, it depends on your research question, the nature of your research, and your goals. For example, if you are interested in how students develop as writers, then you need to look at several of the students' papers. A lot of it boils down, again, to being fair and honest in your assessment—to not manipulating the artifacts you collect so that you only see what you want to see.

CONCLUSIONS

This chapter has addressed three common tools in qualitative research. The first, reading, is one that every good researcher relies on, and the others are tools that most qualitative researchers in writing use to at least some extent. In composition studies, researchers observe classrooms, workplaces, or other settings and they collect documents from participants that they examine, systematically or informally, for certain kinds of evidence. The next chapter addresses two additional tools that you will likely consider using: interviews and surveys. Taken together, all of these tools offer a range of possibilities for obtaining useful and focused information from your qualitative research that addresses your research question.

How Do I Find Answers? Carrying Out Your Qualitative Research Study, Part II— Interviewing and Using Surveys and Questionnaires

This chapter completes our discussion of the research tools available to qualitative researchers by addressing three additional tools: interviews, surveys, and questionnaires. Because our presentation extends from the previous chapter, a general issues section is not included here. However, we do define the tools we present and address when, where, why, and how they are used. Additionally, we again show how particular lenses or biases can affect how you use each of the tools, as well as address the ethical issues that can arise in using them.

Before discussing the specifics for each of the tools, let us briefly explain how they are used by qualitative researchers. First, qualitative researchers interview participants to understand and make meaning of their participants' experiences. Interviews are opportunities to explore with a participant, in an in-depth manner, a situation, experience, or issue. Interviews provide information both about the person being interviewed and from that person's perspective. They provide insight into the person's thoughts, perceptions, feelings, motivations, responses, and actions in relation to the issues or situations being explored in the research.

The other tools discussed in this chapter, surveys and questionnaires, are not always associated with qualitative research; they are included here, however, because we believe they offer researchers the opportunity to expand the scope of what they are able to learn through their research. They do this by providing information quickly from a larger group of people. Qualitative researchers can thus use questionnaires or surveys to gather data on how a larger number of people think about a particular aspect of their research question. This chapter distinguishes between surveys and questionnaires by defining surveys as a more formal tool for gathering information. Researchers will have different reasons for using a survey or questionnaire: They will typically decide which of the two to use based on what they hope to learn and/or, in some cases, on which they feel they are best equipped to use.

ADDITIONAL METHODOLOGICAL TOOLS FOR RESEARCHING YOUR QUESTION

Interviews

As a qualitative researcher, you will probably carry out at least some interviews during your research, and interviews may even end up being a primary methodological tool for you. Whom you interview, what you ask them, and how often you interview them will depend on your research question and on your goals for your research. These decisions may also depend on where you are in the research process and how much time you and your interviewees have for the interview. Because of these and other variables in interviewing, there is no single, correct way to carry out an interview. There are, however, a number of considerations in doing interviews that can help make them more effective. These considerations are addressed throughout this section.

Knowing When and Why to Do an Interview

Interviews take many forms (from spontaneous to formal) and can be used for many purposes (to learn about an issue, event, or individual's life story). Their purposes may also vary depending on when they are done during the research process. For example, as we discussed in chapter 2, you might carry out interviews early in your research to obtain preliminary information to help you focus your interests. If you were interested in learning about the revision habits of students in an introductory college composition course, for example, early fact-finding interviews with students could provide important background information for your research. Interviews at this stage of your research can provide you with information about your topic and can give you a sense of the different perspectives that exist on it.

As you carry out your research, you can use interviews as either a primary or a secondary tool for collecting information. Interviews can provide key information about the issues you are researching or about the themes that end up emerging in your research. With the revision study, for example, if you discover that some students turn in their first drafts for a grade without revising, you can interview those students to explore why. Interviews can also shed light on situations and events, and on people's experiences and actions in relation to or in response to those events. As an example, when Ann worked with the three physicists, she interviewed each of them almost weekly to explore why they interacted in the ways they did during the meetings in which they reviewed the paper they were writing. Like Ann, you might have participants whom you interview regularly (even on a set schedule) throughout the duration of your research, and/or you might have individuals whom you interview just once or a few times. In her research, Ann also had special experts whom she interviewed periodically to help clarify the technical aspects of what she was observing in her research.

Depending on the type of study you are doing, you might also interview participants to obtain their personal stories. Many researchers carry out interviews to construct oral or life histories (see, e.g., Gluck & Patai, 1991). These interviews explore an individual's life experiences, usually within a specific culture or place, or under specific circumstances. Debra Brandt (1994, 1995, 2001), for example, has interviewed people from different socioeconomic and educational backgrounds to learn about their life experiences with reading and writing.

You may also use interviews to extend or round out your understandings of an issue toward the end of your research. For example, if you studied revision and concluded from your findings that students do not revise because they dislike the revision process, then you could do follow-up interviews with certain students to explore why this is the case. Such interviews could help you determine whether the students feel they lack strategies for revising, whether revision seems like too much work to them, whether they are anxious about it, or whether the rewards for it seem insufficient.

Understanding and Deciding How to Conduct an Interview

Another consideration with interviews is how to carry them out. Because different kinds of interviews require slightly different approaches, it helps to understand the various kinds you can do. For example, interviews can be spontaneous, informal, or formal, and which kind you choose will depend on several of the following factors:

- Your research question and the goals of your research.
- The stage you are in your research.
- The other research strategies you are using.
- What you hope to accomplish with your interviews, or the role the interviews will play in your research.
- The nature of your research project (whether it's a long-term study, a short-term one, one that involves many participants, or one that only involves a few participants).
- Your interviewees (who they are, what their positions or roles are, how they are positioned in relation to you).
- Even yourself (your role, your relationship with your interviewees, your persona as an interviewer).

Spontaneous Interviews

If you are a participant observer, or if you are doing a study that entails spending a good deal of time in a setting, you may end up interviewing participants spontaneously. Spontaneous interviews occur at just about any point in time and are a lot like everyday conversation. They are essentially ad hoc interactions with your participants, usually in response to, or just after, certain events or observations. For example, you might observe something of note in your research setting and engage one of your participants in a conversation about it. These can be great opportunities to discover new information, especially because participants are likely to be relaxed during these kinds of interactions and because the context for the discussion is immediate. Although you can't prepare for these interactions in the same manner you would for a more formal interview (e.g., with thoughtfully prepared questions), you can look for opportunities for them and be prepared to take advantage of those opportunities when they arise. If your research setting is conducive to these sorts of interactions, we suggest that you keep a notebook or tape recorder with you at all times so you can be ready when they do occur.

Informal Interviews

Informal interviews share some features with spontaneous interviews but usually have more structure. Informal interviews are interviews you plan, but they are still flexible, especially in regard to the questions you ask and how you structure and direct the interview. Usually you write out questions in advance for these interviews, but you do so knowing that you will work through the questions loosely or that you may even end up departing from them. You might also just write down topics as starting points for discussions. The main advantage of informal interviews is their flexibility: You can easily go off in unplanned directions in order to pursue lines of thought that intrigue you. This flexibility, however, demands that you listen well to the person you are interviewing and follow up on what you hear. Kathryn Anderson and Dana Jack (1991) offer a good example of the potential of this kind of interview in "Learning to Listen: Interview Techniques and Analyses." These authors' own experiences with interviewing show the risks researchers take when they don't listen well to or follow up on their participants' comments. Their own failures to pursue potentially productive lines of discussion with their participants illustrate how a researcher's preconceived notions (the researcher's personal or theoretical lenses) or compulsion to stick to a script (e.g., a structured list of questions) can severely limit interviews.

Because of their flexibility, informal interviews are probably the most common types of interviews qualitative researchers use. Ann used this type of interview throughout her work with the physicists, and she did so because it facilitated getting a better understanding of the topics that came up during her interviews. Usually, she started with a list of loosely structured questions that she used to guide and prompt her.

Formal Interviews

As a qualitative researcher, you may also decide to carry out more formal interviews. With formal interviews, you develop a list of well-articulated questions usually arranged in some sequence and you stick to those questions throughout the interview. Whereas choosing this approach depends on your purposes and goals for your research, conducting formal interviews offers one way of maintaining consistency across your interviews. In some cases, it may be important that you ask participants the same questions in exactly the same manner and order. Doing this can also make your findings more consistent because you can be sure to ask each person the same thing. And doing this with multiple participants can make it easier to quantify your findings, which you may also find desirable in some circumstances.[1]

You can also use formal interviews to interview several people at the same time. You might do this because you are interested in discovering how a group of people feels about a particular topic, or because you believe the group dynamic will provide useful information for your research. Of course, there are down sides to such interviews, with one of these being the potential for certain individuals to dominate. Another is the potential for group speak, or people expressing opinions that

---

[1]Some of you may be thinking, "Wait, I decided to do qualitative research so that I wouldn't need to count or quantify my findings." That's certainly legitimate, but it often is useful in qualitative research to count certain findings (e.g., the number of times a term is used, the number of times a particular reference is made, etc.). When you do things in a systematic and uniform manner, this is easier to do, and more legitimate. You can say something like, "In response to this question, X number of participants said Y this number of times."

they believe to be popular with the group. Generally, however, group interviews can offer a useful way to obtain multiple perspectives simultaneously.

One common type of group interview is a focus group. In these you gather a group of individuals around a common issue or topic, and the group discusses that issue or topic. As with interviews, these can be conducted in a more or less formal manner depending on the purposes of the researcher, the nature of the discussion, and so forth. Many of our own students have relied on focus groups for their research. For example, one student who was interested in safety information on the Web planned to follow up her online survey with focus groups. She would ask for volunteers at the end of the survey and then follow up and contact those volunteers to develop the groups, where she would ask for additional information about the presentation of safety information on the Web and about their preferences for both what and how it's presented. Another student followed up his online surveys of employees at his company with focus groups in order to delve more deeply into the patterns he was seeing in their survey responses. He was interested in the employees' responses to the various change communication strategies that the company was using.

In-Depth Phenomenological Interviews

A final kind of interviewing we would like to discuss comes from Irving Seidman's *Interviewing as Qualitative Research* (1998). Seidman defines "in-depth, phenomenological interviewing" as an approach to interviewing designed "to have the participant reconstruct his or her experience with the topic under study" (p. 9). His detailed and fairly specific approach is based on a three-interview series with a single person. The first interview is a focused life history in which the interviewer asks the interviewee questions designed to put the person's experience in context (asking "how?" instead of "why?"). In the second interview, the interviewer asks the interviewee to reconstruct her experience on the topic under study in concrete and detailed fashion. In the third interview, the interviewer asks the interviewee to reflect on the meaning of the experience.

Next, Seidman lays out a step-by-step approach for analyzing the interview series that entails creating two kinds of products: individual profiles of those interviewed and a thematic study of the interviews. He calls this approach shaping "the material into a form in which it can be shared or displayed" (p. 101).

He asks the researcher, first, to read and mark the transcript with labels, focusing on those passages that seem interesting. Next, he suggests making two copies of the marked text. Then he tells the researcher to literally cut up one of the marked copies, either with scissors or on the computer, and to file the pieces into folders (either in manila folders or in folders on the computer) according to the labels that the researcher has devised. These folders are to be used later for the thematic study of the interviews.

The fourth step entails taking the other copy of the transcript, selecting all the marked passages, and putting them together into a single transcript. The goal of this step is to reduce the length of the original transcripts by a third to a half. Next, he suggests that the researcher read the transcripts again and underline the most compelling passages. At this point, the researcher can craft a narrative based on the passages, using the first-person voice of the interviewee. Another kind of product he encourages is a thematic analysis of the interviews, which can be used to enhance the profiles (see the section on thematic analysis in chap. 6).

According to Seidman, what is valuable about this approach to interviewing is the dialectical relationship the researcher establishes with the material provided by the interviewee. One student with whom we worked liked this approach for two reasons: It forced her to really focus in-depth on the words of the young woman she interviewed, and it allowed her to preserve the young woman's voice, something we say more about in chapter 7.

Preparing for and Carrying Out an Interview

We both have used interviews extensively in our qualitative research projects. Both of us have carried out successful and unsuccessful interviews, and we have also had experience being interviewed by others, some of whom were well-prepared and asked good questions, and some of whom were not well-prepared at all. What we have come to understand through all of these experiences is this: Although interviews may seem to be a fairly easy and accessible research tool, they—like all research tools—require a thoughtful approach and good preparation. In other words, there are ways to do interviews that, with practice and experience, you can learn and keep improving upon. We both have certainly made mistakes in carrying out interviews, and we suspect that most people who have done them and reflected at all on what they have done would admit the same. This section addresses some important concerns you should have as you prepare for and then carry out interviews.

Doing Your Homework

Preparing for an interview entails a number of tasks, not the least of which is doing your homework. In your observations for the first prompt, you probably noticed that at least some of the interviewers had read something written by or about the person they interviewed, and/or had done research on the topic for the interview. That is common, and something you should do as well, both for formal and even for informal interviews. If the individuals you are interviewing have authored

Prompt 1: *Evaluating Interviews.*

Listen to and/or watch three different interviewers at work. We suggest that you select three different kinds of interviewers; for example, a shock-talk host like Jerry Springer or Montell Williams, a more mainstream popular interviewer like Oprah Winfrey or Larry King, and a more "serious" or specialized interviewer like Bill Moyers or Terry Gross. You might also select any of the hosts of the network or cable news programs. When you watch these interviewers, look beyond the obvious differences in the content of the interviews and the characteristics of the interviewees and ask these questions:

- What kinds of questions do they ask?
- How did they prepare for the interview (e.g., what have they read or what do they seem to know already)?
- What kind of presence or persona do they have as interviewers? How do they conduct and carry themselves? How do they act? How does their ethos come across?
- How do the interviewers interact with and draw out their interviewees?
- What qualities strike you about the interviewers?
- Which interviewer do you like the best and for what reasons?

---

articles or books, then you need to read or at least familiarize yourself with what they have written, as well as familiarize yourself with what has been written about the individuals. You may also want to conduct some general background research (e.g., find out about the person's education, work, personality, achievements, and interests). Some useful resources for this background work are company publications (e.g., newsletters), books, the Internet, periodicals, and newspapers. Even with spontaneous interviews, some preparation is important. Continually reviewing the information you have already collected for your research—and knowing as much as you can about your participants—can help you make the most of your interviews, whether they are formal, informal, or spontaneous.

This sort of background work initially requires you to invest some time, but ultimately it saves time. Specifically, it will save you time during the interview because you won't need to ask about those things you already know, and because your participant won't have to explain those things. Time is usually limited during an interview, and if you are interviewing a person only once, there is even greater pressure to make good use of your time. This doesn't mean that you need to spend weeks preparing for the interview, but it does mean you should spend at least some time brushing up.

Advance preparation is also a way to show respect for and build rapport with the person you are interviewing. Knowing the individual's area of expertise can make a strong first impression on the person; it can also help break the ice, putting both of you at ease. Additionally, this kind of preparation can fill in gaps during the interview and can lead to a more fruitful exchange. Your knowledge can also help you ask better questions. As examples of the importance of preparation, we each share one of our own experiences in the sidebar for this section.

Sidebar 1: *Preparing for Interviews*

Ann: When I did my research with the group of physicists, I initially looked at their publications in two related journals: *Physical Review* and *Physical Review Letters*. I wanted to see how each of these journals functioned in the field, and how authors wrote articles for them (one is a rapid dissemination letters journal and the other a more traditional scientific journal that publishes research articles). As I carried out this research, an opportunity arose for me to interview the just-retired editor-in-chief of these journals, which were both published by the American Physical Society. To prepare for this interview, I read biographical information about this individual, and I also reviewed his publications. I talked to other physicists who knew him, and I did some background research on the journals and on the American Physical Society. These were pre-Internet days, so I spent a good deal of time in the library, and I also joined and then had information mailed to me by the APS. By the time I carried out my interview, I knew a substantial amount about the person I was interviewing, and I was familiar with both the APS and the two journals. I asked questions that built on my knowledge and that elicited unique information from my interviewee, information that wasn't readily available in other sources. My interview lasted 2 hours (I had planned for an hour, but he graciously offered me additional time because he enjoyed the exchange), and it yielded many valuable insights (e.g., on how these publications were functioning in the field and on how they would likely change). Our time together was productive for both of us, which I attribute in large part to the preparation I did.

Cathy: When I started my study of community organizing, looking at its connections to getting out the word on teachers' work, I had to enter into fields of study about which I knew very little. My goal was to learn about community organizing from those who engaged in this activity in various disciplines. Thus, I realized that I needed to know something about public health, environmental activism, and progressive political science. And while I knew something about many of these fields—enough to have a dinner table conversation with someone about their work—I also knew I didn't know enough to make my 1- to 2-hour interviews with the various community organizers valuable. I approached this lack of knowledge in two ways. First, I did a lot of reading on my own, selecting books whose titles seemed right and whose authors were somewhat familiar. I also asked some of my interviewees, prior to our interviews, to recommend a book or article that might be useful for me to read before we got together. I found this second strategy to be very helpful. The interviewees and I could then begin the interviews with some common language; additionally, the interviewees seemed to appreciate my seriousness of purpose.

Prompt 2: **Doing Homework.**

Think again about the interviews you observed for the previous prompt. Focus on the types of preparation or homework the interviewers might have done. What difference did this preparation, or lack of it, seem to make in their interviews?

The information you obtain about the person you are interviewing will become part of the larger context for your interview. That context also includes the topic of your research, your research question, the purpose of your research, your participants, and the subject of the interview. This context will also inevitably affect the questions you ask.

Developing Your Interview Questions

Developing questions is another important task in preparing for interviews. Because certain kinds of questions will elicit more and better information than others, it is important to devote time to this task as well. The best questions are those that get your participants talking, revealing information that hasn't been revealed by any other source. Problematic questions are those that lead either to perfunctory responses or to responses that are simply what you as the researcher are looking for. The kinds of questions you write, then, are key: Your goal should be to write questions that will encourage honest, thoughtful, and full responses from the individuals you interview.

Prompt 3: **Writing Interview Questions.**

Select someone to interview, preferably someone connected to your research, and do some homework for the interview. Next, begin writing questions or prompts for the interview (decide what type of interview you will carry out and for what reasons and then write your questions accordingly). As you are developing your questions, brainstorm, first individually and then with the class, some characteristics of good interview questions.

In generating and discussing characteristics of good interview questions, you might have identified some of the following traits of good questions:

- They are open (requiring more than a yes–no response, and/or not suggesting a particular response).
- They are clear and uncomplicated (meaning, among other things, that you don't have multiple questions or potential meanings embedded into a single question).

- They are focused (e.g., on the purposes of your inquiry and the subject of the interview).

When you write questions, you may also want to consider how you sequence them. For example, it is common to ask more foundational and general questions first because these kinds of questions reassure interviewees and put them at ease. They also establish a common ground. Your later questions can then be more specific and more focused on what you hope to learn from the interview.

Prompt 4: *Revising Your Questions.*

After a few days, consider again the questions you developed for your interview for the previous prompt. What are some ways in which you might revise or improve your questions? Revise your questions, noting those things that you change and your reasons for changing them.

Setting Up the Interview

In addition to doing your homework and developing your questions, you also need to set up the interview, which entails thinking about the following:

- When you will conduct the interview (what day of the week and time of day).
- How long it will last.
- Where you will conduct it.
- Whether you will interview that person again and/or whether you will have access to the individual after the interview.

Although seemingly basic, all of these are things that can be easily overlooked or taken for granted. Later we address some considerations connected to each of these issues.

First, timing is important in interviews because it can influence a person's responses (e.g., Friday afternoons, or even the end of any work day, probably are not good times for doing interviews). Location is also important. Will the interview be on your turf or theirs? Or, will it be at some neutral location? Every location has advantages and disadvantages. For example, people usually feel more comfortable in their own space, and there may also be good reasons for using that space (e.g., if you are doing a life history, or if what you are interviewing the person about is closely tied to what they do and where they do it). The interviewee's own space, however, can also have drawbacks (e.g., there may be too many distractions). Your space can have similar problems. And, if you choose a neutral setting, which may be most ideal, then you should choose a setting that will be quiet and private enough for an interview. It is difficult, for example, to carry out an interview in a coffee shop or restaurant. These settings often seem appealing, but they are usually too noisy and have too many distractions.

You should also consider how long you will need for the interview, and make sure the amount of time you allocate is acceptable to your interviewee. Most research interviews last no more than an hour, although some types of interviews (e.g., life histories) need to be longer. If you are respectful of your interviewees' time, they may offer you more of it; however, you should never count on that. In other words, always plan for the time your interviewee agrees to. If you are new to interviewing, and lack experience with this, prioritize your questions so that if you do run out of time, you at least will have covered the important ones. You can also sometimes ask interviewers for follow-up sessions and hope they will agree to them.

Conducting and Transcribing the Interview

What all of the planning you do for interviews really boils down to is making your interviews worthwhile and productive experiences both for you and for the people you are interviewing. But, there are additional considerations that will arise when you carry out your interviews. For example, because we encourage you to think of interviewing primarily as an interaction between you and the interviewee, both parties must be attentive and responsive. As the interviewer, you set the tone for the interview. Interviewers, therefore, need to listen carefully so that they hear and understand what the interviewee is saying. Interviewers also need to guard against butting in and not giving participants the opportunity to talk. They need to guard, as well, against getting distracted by their own thoughts or by their concerns with how they are going to respond or what they are going to say next. Finally, interviewers need to be careful about filtering what they hear through their own lenses, which is discussed later.

Sometimes, interviewees may not hold up their end of the interaction, seeming to be uninterested, reticent, or even overly eager. As an interviewer, you have to be aware of and responsive both to your interviewees' verbal responses as well as to their nonverbal cues, including body language. You need to respond, adapt, and make decisions throughout the interviewing process, which sometimes will require you to stray from your original questions in order to keep the conversation alive and focused. For example, if the person says something interesting but not directly related to your research, do you follow up on it or gloss over it? What you do in this situation will depend on a number of factors, including the role and stance you have adopted as an interviewer and the information you hope to obtain. You should ask yourself these questions: What might you learn if you let the interviewee go on in a different direction? Would that knowledge be useful? If you need to get the person back on track, what are the best ways to do that? Thinking about possible interview dilemmas in advance can be helpful, but you also need to reflect constantly on your approach to and experiences with your interviews. Remember, you have a great deal of control over your interviews; however, you also need to be careful: You don't want to exert so much control that you limit the process.

Prompt 5: *Reflecting on What Makes Interviews Effective*

Think again about the interviews you observed. What did you like best about them and why? What did the interviewers do that seemed effective? What did they do that seemed ineffective? How would you characterize the personal styles of each of the interviewers? What were some of their positive and negative attributes?

Finally, you also need to make decisions prior to your interviews about how you will take notes—whether you will write out your notes by hand, record them, or both. Whereas technology provides an accurate record of what was said, it isn't always reliable. Thus, we recommend that you combine writing and recording. Whether you record your interviews by hand or record them digitally or with a tape recorder, you should review and transcribe them (or fill them in) as soon as possible after the interview. Doing so allows you to draw on your memory of what was said, especially if there are any gaps in your notes or places in your recording that are not audible.

We realize that transcription can seem like an onerous and time-consuming task. We also realize that researchers, with all they have to do, may be tempted to skip this stage all together, relying instead on just listening to the recorded interview. In our experience, and that of our students, foregoing the transcription stage is a real mistake. The nuances and connections that become clear in reading the transcription of an interview are extremely valuable. These nuances and connections truly cannot be matched by listening to just portions of recorded responses.

You also have several options when you transcribe.[2] One is to transcribe everything that was said so you have a verbatim record of it, and another is to transcribe only selectively. If you do the latter, you can transcribe those parts that seem most relevant while simply marking other places, making note of what those sections address. This approach is especially useful if you don't need a complete record of your interviews, and/or if other constraints prevent you from transcribing everything.

As you review and transcribe your interviews, you should also make note of what you observed during them (e.g., the body language you observed). Specifically, if you

---

[2]In addition to the options you have in deciding how much of an interview to transcribe, you also, if you use a traditional tape recorder, have options in terms of the equipment you can use to transcribe. Both of us, for example, transcribe using hand-held portable tape recorders and head phones. If you use these, be sure the recorder has a pause button. You can also purchase or borrow a transcription machine. These have foot pedals that allow you to keep your hands free while stopping and backing up the tape.

wrote down notes to yourself during the interview, elaborate them before you forget what you meant. (At this time, you may wish to review the different note taking strategies and types of research notebooks that were described in the previous chapter.)

Prompt 6: *Carrying Out and Transcribing an Interview.*

Arrange and carry out an interview. Transcribe either a portion or all of it and review the transcription. What do you notice in reading the transcribed text that you think you might have missed in the actual interview? Share with your classmates your experiences transcribing the interview and reviewing the transcription.

The Role of Lenses in Interviews

As they did with your observations, your personal and theoretical lenses will have an impact on the interviews you do. How you view things, including the biases you hold, may even influence whom you decide to interview. You may end up deciding, for example, that some people are more important to interview than others, which may lead you to overlook potentially valuable sources. Your personal and theoretical lenses will also influence the questions you ask, as well as how you interpret and make meaning of your findings.

Personal and theoretical lenses strongly influence how researchers carry out interviews. They can influence the entire tone of an interview, including how the interviewer views, approaches, interacts with, and perceives the interviewee. Many theoretical perspectives, like feminism and social constructionism, for example, encourage viewing participants as co-creators of meaning and even as collaborators in the research process. Good examples of this are offered in Gluck and Patai's *Women's Words: The Feminist Practice of Oral History* (1991), especially in Katherine Borland's contribution on interpretive conflict in oral narrative research. In this piece, Borland's participant, her grandmother, took issue with Borland's interpretation of her experience. Borland used this experience to reflect on the issue of interpretive authority in research. She concluded that "by extending the conversation we initiate while collecting oral narratives to the later stage of interpretation, we might more sensitively negotiate issues of interpretive authority in our research" (p. 73). Other perspectives encourage researchers to retain authority and control over the interview process. As with all aspects of research, we encourage you to identify, reflect on, and acknowledge your own personal and theoretical perspectives.

Interviewing Ethically

Interviews, like all of the tools addressed here, entail numerous ethical considerations. You should approach all the interviews you do with integrity and honesty, and you

Prompt 7: *Reflecting on How Your Own Lenses Influence Your Interview.*

Reflect on the interview you carried out. Think about whom you chose to interview, what questions you asked them, how you asked them, and how you interpreted what you heard. How might your own lenses have influenced those decisions? Try to think of at least one example of how a particular viewpoint you hold influenced something you did either during or after the interview.

should always show respect for the individuals you interview. You should also be aware of how the decisions you make prior to, during, and after your interviews have ethical implications. Even the issue of whom you decide to interview—or not to interview—may be an ethical one.[3] You might, for example, consciously or even unconsciously choose not to interview certain individuals on the basis of how you feel about them personally and/or on the basis of what you think they might contribute (or not contribute) to your research. We all have personal preferences, so we need to be aware of those preferences and of how they might influence our research.

The manner in which you carry out your interviews—for example, what you ask or do not ask your interviewees or what you choose to pursue or not pursue—will also raise ethical issues. You may limit your findings by choosing not to pursue an issue that runs counter to a perspective you hold. How you ask questions also has ethical implications. We talked previously about questions that suggest certain responses—leading questions. If you use these, you curtail your interviewee's ability to respond openly. You also need to guard against making comments that finish your interviewee's statements. If you do too much of this, you limit your findings: Your findings will end up reflecting mostly what you think and believe, rather than what the person you interviewed thinks and believes. Because the point of interviewing is to capture the perspectives of your participants, speaking for them too much will fail to accomplish this.[4]

---

[3]In many qualitative studies, deciding whom to interview is fairly easy; however, in some studies you need to give serious thought to this issue. Depending on the type of study you are doing, you may also need to think about how many people you should interview, and even how many times you should interview them. These decisions become ethical ones when they have implications for your findings and/or their reliability. For example, if you are trying to discern patterns or similarities among the individuals you are interviewing, you will likely need to interview enough people to make those discernments. You may also need to consider whom you will interview (e.g., people from different socioeconomic backgrounds, different educational backgrounds, different generations). Your concern is with obtaining reliable information given what you intend to do with the information—the claims you hope to make, what you hope to show, and so on.

[4]It is important in interviewing, and really in using any of the methods we discuss, to guard against using the tool to prove your own point, whether that entails asking only certain kinds of questions, pursuing only certain issues, leading your interviewee to make certain comments, and so forth. Unethical interviewers might even manipulate or deceive the person being interviewed in order to achieve their own purposes.

Prompt 8: *Considering the Extent to Which You Take Control of an Interview.*

Review again your transcription of the interview you carried out, this time highlighting all of the places where you did the talking. Look at these and see what you are saying in relation to what your interviewee is saying, and how frequently you are saying it. Are you finishing statements for the interviewee? Are you suggesting meanings for the statements the individual makes? If you are seeing a lot of yourself in the interview, then you may want to consider how to make the voices of your interviewees more prominent.

Ethics also enter into other aspects of interviewing, such as how you interact with the people you interview—for example, how you handle your own and their voices, how you treat them, the kind of rapport you establish with them, and even the overall relationship you have with them. It is not uncommon for researchers and participants to become friends in qualitative research, especially if they work together over a long period of time. However, research friendships are not without risk: They can lead to bias, and they can prevent you from seeing things you might otherwise see. They can also cause unanticipated tensions that can influence your research.

Finally, you should also be concerned with how you interpret and use the information you obtain from interviews. For example, it is not uncommon in interviews for a person to tell you something that is pertinent to your research and then request that you not use the information. If your interviewees trust you, they may share personal information with you that they wouldn't share with the public. Some of this information, not surprisingly, might be relevant to your research. Interviewees might also tell you something that they want you to use but that you don't find relevant or even appropriate. These kinds of issues arise primarily when you begin writing up your findings, so they are discussed again in chapter 7. It is then that you need to decide what to use and not use, and it is then that you may end up struggling with whether or not to use some information that implicates your interviewees or that portrays them in an unfavorable light. There are no easy answers to these issues. And you may not even become fully aware of them until you encounter them firsthand, which is our experience. Being aware of them, however, can help you when and if you do encounter them.

Surveys and Questionnaires

What They Are and When You Use Them

Surveys and questionnaires can serve a number of purposes. Writing researchers often use them to gauge students' or other individuals' perceptions of or understanding of a subject. For example, one of our students surveyed her own students at the beginning of the year to see how much out-of-school reading they said they did and how they selected the books they read. She surveyed them again at the end of the year to see if they had changed at all in this regard (see appendix A). Surveys also reveal people's responses to particular situations. Another of our students, for example, surveyed employees at his company to see how they felt about the different change communication strategies the company was using. Surveys, further, can reveal something about people's perceptions, behaviors, habits, and practices, along with their reasons for those. For example, still another student carried out an extensive web-based survey to find out whether and how people look for and use safety information on the Web (this student's research proposal is included as an example in chap. 3).

Many researchers use the terms *survey* and *questionnaire* interchangeably, but we view them as being slightly different. Questionnaires, in our view, are less formal than surveys; they generally require less elaborate preparation and are often more open-ended in design. They tend to have a slightly different purpose as well. Whereas surveys are generally intended to gather responses from a large number of people, often disparate and even unfamiliar to the researcher, questionnaires tend to be targeted at a smaller number of people and situated in a particular context or situation (e.g., the researcher's work context). For example, Ann used questionnaires for several successive semesters to find out how students in her classes felt about their experiences completing workplace projects with clients. (See the sidebar in this section for examples of these questions.) The two of us also used questionnaires with the graduate students in our research classes over the past several years as we planned and used various drafts of this book. These were concerned primarily with discovering how well the different sections of the book worked for them as beginning researchers. We asked about their sense of its organization, about the tone and voice of the writing, about our explanations of strategies and our discussions of lenses and ethics, and about the prompts. We also asked them for ideas for topics we might not have thought to include.

Perhaps the best way to think about the differences between surveys and questionnaires is to imagine a continuum with informal questionnaires at one end and formal surveys at the other. We devote much of our discussion here to the construction of more formal surveys; however, we encourage you to consider how you might adapt the strategies we present for constructing surveys if your intent is to produce something less formal. Formal surveys can yield important and highly relevant information; however, so can less formal questionnaires. To decide which might be best suited for your research, think carefully about your research question, the context for what you are doing, the numbers and kinds of people you hope to reach, their and your own time constraints, and the kinds of information you wish to discover.

Sidebar 2: *Sample Interview Questions*

Questions for Students Participating in Client Projects

- What are your perceptions of who your audiences were for these projects? Who were you writing to and for?
- What difficulties did you have addressing these various audiences?
- Did the projects seem real to you? In other words, did they seem like real-world, workplace projects or not? What seemed real and what didn't? Or how did they seem real, or not real?
- Did the projects seem instructionally useful? In what ways? How could they have been more useful instructionally?
- What would have made your experience with the projects better?
- Was your motivation level higher or different from that with regular class projects?
- What are your perceptions of the level to which the clients valued the projects?
- What, generally, did you like about the projects?
- What, generally, did you dislike about the projects?
- Are the projects worth doing? Why or why not?

Prompt 9: *Comparing Surveys and Questionnaires.*

Use our definitions to locate examples of both a survey and a questionnaire. Compare and contrast the two and list the features they seem to share and those that distinguish them. Next, think of a reason for using a questionnaire or carrying out a survey in your own research. Decide which would be most appropriate and explain why. Also consider what you might hope to learn from whichever of these tools you select. Given what you hope to learn, what are some of the questions you might ask and how would you word and present them?

Whether you decide finally on a survey or questionnaire, keep in mind that these tools usually do not provide the same depth of information that interviews do. They can, however, be effective precursors or follow-ups to interviews. Our student who was studying change communication, for example, followed up his surveys with interviews in order to probe the issues he found in the surveys. The student who was interested in safety information on the Web planned to follow up her survey with focus groups. But surveys can also be valuable as follow-ups to interviews. For example, looking over the responses of several interviewees might lead to important questions that you use in a survey or questionnaire in order to gather responses from a broader audience. Again, the main advantage of surveys is their breadth. You can collect a lot of information in a short amount of time. And because they are usually anonymous, respondents may be more willing to answer them truthfully, especially if the subject is sensitive.

How to Construct Surveys and Questionnaires

Consulting With Experts

Developing an effective and reliable survey or questionnaire requires a good deal of skill. In fact, many researchers even shy away from these tools because of how complex they can be. However, we encourage you to assess honestly whether a formal survey, which is the more difficult of the two to construct, would be an effective research tool for you and, if so, to do whatever you can to construct one that will be reliable. This section offers advice for doing that. At the same time, we strongly encourage you, if you do decide to construct a survey, or even a questionnaire, to consult other resources (e.g., books on surveys and/or individuals who are expert in survey design). Many universities offer free services for people interested in developing and administering surveys. You should consult your campus research office to determine if such services are available and who offers them. Then, once you identify someone who can assist you, work closely with that individual as you develop, test, and administer your survey.

Prompt 10: Consulting Experts.

If you are reading this text as part of a course, choose someone from the class to look into your university's research services. A couple of students might even be assigned this task because your university might have a variety of services. You should try to find out what services are available and who provides them. Others should locate additional resources that address surveys (e.g., books or Web sites).

Planning a Survey or Questionnaire

Whether you administer a formal survey or a less formal questionnaire, you need to consider a number of factors as you plan it. The first and most significant of these are audience and purpose. In other words, whom are you targeting with the survey or questionnaire and what do you hope to achieve with it? From our own experiences with these tools, we can honestly say that audience and purpose are the easiest things to lose track of, but they are also the most essential in making surveys effective. For example, if your purpose is to determine perspectives on new technologies, you would potentially construct one kind of survey for an audience of people over the age of 60 and another kind for an audience of people under 30 (e.g., you might do one as a paper survey and the other as an online survey, and you might also ask different kinds of questions). If you are surveying both of these age groups to determine the features of Web sites they prefer, then you might well decide to use an online survey with examples for the respondents to reference. Our student who was concerned with where people look for and what they want from safety information on the Web decided rightly to do an online survey targeting a broad cross-section of individuals. Our student who studied his company's change communication strategies used his corporate intranet, which employees consult regularly, to administer his survey. Audience and purpose are determining factors, then, in how you design a survey: in the kinds of questions you ask, in the length you make the survey (see later), and in how you administer it.

Prompt 11: *Considering Audience and Purpose.*

Write down both the audience and purpose for a survey or questionnaire you might construct. What are some characteristics of your audience, and what do those characteristics, and your purpose, suggest for how you might design your survey (the kinds of questions you might ask, the length you will make it, and how you might administer it)?

Two additional considerations in planning surveys are sample and sample size; in other words, who, or what groups, and how many in those groups will you target? If you are using a less formal questionnaire, you likely will have some flexibility here. However, if you want your results to be statistically significant, you need a sampling strategy and a certain sample size. Some common sampling strategies for surveys include the following:

- Typical case sampling, which gives you a sample that is normal or typical of a population.
- Homogeneous sampling, which gives you similar cases.
- Deviant case sampling, which gives you extreme or unusual cases.
- Maximum variation sampling, which gives you a sampling of a wide variety of cases.
- Convenience sampling, where cases are selected based on convenience (see Glesne, 1999, p. 28).

Once you have a strategy for selecting your sample, you also need to consider how many people you need or want to survey. This entails considering response rates, which for formal surveys, even in the best of circumstances, usually are less than 50% and sometimes even less than 30%. If you are doing a more formal survey and the response rate is a concern, it is advisable to consult a statistician or your survey design expert to get their recommendations for sample size.[5] Our student who surveyed consumers about their need for and concerns with online safety information consulted with a company that provides surveying services, and they told her she needed at least 500 responses for statistically significant results. She ended up sending her survey to as many friends, acquaintances, and business associates as she could. She also used professional listservs and, in all of these cases, she asked recipients to forward her survey to others.

---

[5]If your intent is to use your findings in a more descriptive manner. Sample size may be much less of concern.

Once you have identified what you want to achieve with your survey or questionnaire and who and how many individuals you want or need to respond to it, then you can begin the process of constructing it. As you do this, we advise that you give yourself sufficient time to generate numerous drafts and to obtain and incorporate feedback on the drafts, ideally from potential respondents.

A final concern you should have in planning surveys and questionnaires is length. Length is important because it can influence both response rate and the nature of the responses. Individuals, especially those who don't have a vested interest in your research, may decide not to respond if the survey or questionnaire seems too long and if it seems like it's going to require too much of their time. On the other hand, you might have a valid reason, and the right sample, for a longer survey or questionnaire. Generally, however, the simpler and quicker the instrument is to respond to, the better, and the higher the response rate.

Developing Questions

It is likely that you will spend most of your time in developing a survey or questionnaire on writing your questions. They need, first and foremost, to solicit the information you are seeking. In other words, they need to be written so that they end up giving you what you are hoping to get. They also need to be constructed so that they are unambiguous and so that they don't end up deceiving respondents. Similar to interviewing, you want questions that are not biased and that do not lead respondents or evoke unfavorable or negative reactions from them. Additionally, you want questions that provide usable data (e.g., with more formal surveys you may want questions that allow you to enter the responses easily into a database so you can analyze the responses statistically or quantitatively). You may also want to include some redundancy in your questions so you can verify or cross-check responses. Most researchers end up spending a good deal of time on developing their questions, realizing, among other things, that how the questions are worded will have a significant impact on respondents' answers.

There are numerous ways to write questions depending, to a large extent, on the formality of what you are doing. You might have yes–no or true–false questions; multiple choice questions with the option for either a single response or multiple responses (e.g., lists of items with the stipulation that respondents check all that apply); questions that require a ranking of responses (e.g., from most to least important); open-ended and/or short answer questions; and questions that use a Likert scale for the response (e.g., a scale measuring strength of agreement[6]). Most surveys and questionnaires have a combination of question types. (See, e.g., in appendix B, a survey Ann constructed for our university's the Writing Across the Curriculum Program, WAC.)

---

[6]http://www.isixsigma.com/dictionary/Likert_Scale–588.htm

In one of the early drafts of the WAC survey, Ann and her colleagues asked numerous open-ended questions. In later drafts, they formulated questions that would allow them to enter the responses easily into a database and to tabulate them using the statistical software, SPSS. One of the lessons Ann learned from the experience of developing this survey is that it's easy to get mired in the process. In other words, the further into it you get and the more drafts of the survey or questionnaire you produce, the more difficult it can become to discern what's really effective and what might be confusing to respondents. Again, testing is important, but even that has limits (it can be quite useful, but it may not reveal every problem). The claims you make with your results will depend, ultimately, on the effectiveness of your questions and the reliability of the responses.

Prompt 12: Constructing Questions.

Begin constructing some questions for a survey or questionnaire you might want to use in your research. Use the information in this section, along with your considerations of your purpose and audience, to develop your questions. Then ask a classmate or acquaintance to review and comment on them. Finally, if you are able, ask someone who is similar to your prospective respondents to answer your questions.

Deciding on How to Disseminate a Survey or Questionnaire

While you are constructing your survey or questionnaire, you also need to develop plans for disseminating it. One of the main choices you need to make is whether to do an online or paper survey. The Internet offers a very useful tool for administering surveys, especially formal surveys that are intended to reach a broad audience. A number of researchers have come to prefer online surveys because of the advantages they offer. These include the following:

- Reach—online surveys allow you to reach a larger and more dispersed group of respondents easily and quickly.
- Cost—online surveys save on postage and reproduction costs.
- Ease of response—these surveys usually entail only minimal keystrokes for responding to and submitting them.

Online surveys, however, also can have drawbacks. Some researchers make them too long and complicated, and some fail to consider the technology requirements vis-à-vis the technology their respondents have available. If the survey takes too long to download or move through, if it isn't compatible with the browsers or operating systems respondents have, or if it's confusing and lacks sufficient directional cues, then it may end up frustrating respondents and limiting their responses.

If you are doing an online survey, you also need to consider to whom you will send it and how you will send it (e.g., if you will distribute a URL via e-mail, via a paper letter, or through a listserv). If you decide to send the survey as an attachment, then what measures will you take to make sure respondents can open it, and how will you guard against viruses? What will you do if respondents either don't get or can't open the attachment? Conversely, if your survey is on paper, how will you produce and distribute it, and what will it cost you? Also, how will respondents get the survey back to you? Will you include a self-addressed, stamped envelope or ask them to pay for the postage? You need to consider not only your respondents here but also your budget.

A distinct advantage of online surveys is that they can be constructed so that the responses can be entered automatically into a database. Other surveys require that you input the responses by hand, which can be quite tedious. If you do an online survey, you also need to consider the programming requirements for what you hope to do. If you lack the skills to do these things yourself, you may need to hire someone who has them, and/or you may want to use a service (more of these are becoming available, so you also need to do your homework and select one that will do what you want it to do and at the right cost). Our student who did the web survey, for example, used a service available through Yahoo to author and upload her survey. As web editors become more sophisticated, they too will make the task of constructing an online survey simpler. And your university may also subscribe to a service that you can use, although not all of these are of the highest quality. Sophisticated online surveys have a good deal of flexibility, including the ability to skip automatically to certain questions based on responses to previous ones.

Finally, with online or paper surveys, how you communicate with respondents can make a significant difference in how successful the survey is. Usually it's a good idea to let respondents know what to expect. For example, with online surveys respondents may not be able to tell what the survey entails (usually respondents can't view the entire survey and they often can't move forward until they have begun responding to questions). Ann had this concern with the early drafts of the survey on safety information, which had little information about the scope of the survey and what to expect from it. With online surveys, in particular, you need to provide sufficient navigational cues. Make sure respondents know where they are at and where they are going next in the survey. And, if possible, allow them to back track. These principles apply to paper too, but it is a bit easier to achieve there. Essentially, you want to design a survey that your respondents can complete with ease.

Again, we can't stress enough the importance of involving others as you engage in the process of constructing a survey or questionnaire, whether it's paper, electronic, formal, or informal. Get their advice, feedback, ideas, and so on. New technologies have given us a lot of additional options for administering surveys and questionnaires, but they can't do the work for us. You need to grasp and embrace the technology, or consult with someone who does, to make maximum use of it.

Motivating Respondents

A final factor to consider in constructing a survey or questionnaire is how to motivate your respondents. It may be that your respondents have an inherent interest in the survey and are motivated naturally to complete it. Ann, for example, responded to an employee wellness survey because of her interest in the topic, and Cathy responded to a survey addressing air quality in the building in which her office is located also because of her investment in the topic. What's more challenging is hooking respondents who aren't naturally motivated. One way to do this is to make sure that the purpose of the survey is clear: Let your respondents know how their participation will help answer particular questions, let them know how long it will take them to complete the survey, and let them know how they can learn about the results of your research.

How Personal and Theoretical Lenses Influence Surveys and Questionnaires

Your personal and theoretical lenses will influence surveys and questionnaires in much the same manner that they influence interviews. For example, they influence who you survey and what you ask them, as well as what you do with and how you interpret their responses. Your lenses may also be a factor in whether you even choose to use a survey or questionnaire, what kind of survey or questionnaire you select, and what role you assign this tool in your research. Probably most importantly, your lenses influence the questions you ask and how you formulate them. Here's where your biases can become especially pronounced. In fact, it may be difficult to write questions that do not in some way reflect your biases. Again, asking others for feedback on your questions and testing drafts of your survey or questionnaire are your best strategies for identifying bias.

Personal and theoretical lenses permeate every aspect of the methodological tools available in qualitative research—everything from which ones you select to how you use them and what they end up yielding, and even what you do with what they yield. What is most important is being aware of your lenses so that they don't limit or bias your research. What most of us need to do is find ways to distance ourselves periodically from our research so we can examine our lenses and see how they might be influencing us. It is also good to have someone to check us so we don't lose sight of our lenses and biases. By its very nature, research is a theorized practice. It is a practice, therefore, that requires self-awareness and constant reflection.

Conducting Surveys and Questionnaires Ethically

If your biases are too strong, then your surveys or questionnaires can become unethical. You might end up using a survey more for your own purposes than as an instrument designed to solicit honest responses from participants on issues related to your research. As with all of the research tools, surveys and questionnaires require us to be honest and forthcoming with our participants. That may entail simply telling participants what a survey or questionnaire is about, why we are doing it, and what we intend to do with the results from it. Our participants have a right to this information, and a right, as well, to decide whether to complete the survey. After all, they are giving of their time to complete the survey, so it is reasonable to expect that they will want to be informed about it.

Ethics also come into play in relation to what you do with the information you obtain from surveys and questionnaires. In reporting that information, for example, it is important to be truthful about the sample, its size, and your response rate. If you administered an informal questionnaire with a small number of people, you would report the results much differently than if you administered a formal survey to a large group. Fudging any of the aspects of what you do can seriously compromise the integrity of your findings. You shouldn't claim significance, for example, if you haven't truly achieved it. It is easy enough to qualify your claims and to be honest about the limitations of your data. It is important, as well, to avoid manipulating surveys or questionnaires, or the responses you receive from them, simply to achieve some purpose you have for your research.

Also, if certain questions end up being ambiguous, then you should be careful about how you interpret, tabulate, and report those. As an example, for one of the questions on the WAC survey, we failed to explain all of the options for responding and, therefore, ended up with multiple kinds of responses. As a result, we ended up coding the question based simply on whether or not respondents had marked any of the options (if they had, we coded it as 1 or yes; if they had not, we coded it as 0 or no). By doing this, we salvaged the question. There may also be times when you need to throw out a question all together because it does not yield useful or reliable information. These, ultimately, are judgments that researchers need to make when they see the results of a survey.

CONCLUSIONS

As a qualitative researcher, you have many tools at your disposal for carrying out your research, and many options for using those tools. These two chapters have introduced you to those tools so that you can make informed decisions about which ones to use and so you can get a solid start on your research. However, we have really only scratched the surface, especially given that most of these tools have entire books devoted to them. We recommend that you familiarize yourself with these resources because the more you know and understand about the various tools, the better able you will be to select and use them productively and ethically.

This may also be a good place to talk about the natural ebbs and flows of research. When you are carrying out your research, actually using some of the tools we have discussed, there will be times when everything is going well and you love what you are doing (times when you are learning all sorts of new things that intrigue you), and also times when you lose interest in or end up doubting the value of what you are doing. You might have difficulty finding things to read, experience tensions with your participants, or even find yourself struggling with the "So what?" question. When these occasions arise, our advice is to hang tight or to turn to others (like your advisor) who can encourage and reassure you. You will get through it. If you are part of a cohort group, take advantage of it. Also, be sure to pace yourself (take breaks and re-charge every once in awhile); read new things; talk to others (peers, faculty, professional colleagues); set milestones and have realistic expectations in achieving them; and do something like Julia Cameron (1992) suggests: Set writing dates with yourself by going someplace regularly to just write or think about your work. There are a lot of practical aspects to carrying out research. It is a human endeavor, and every one of us has different styles for and approaches to it. The key is to find what works best for you, whether that's setting aside a certain amount of time each day to complete a task or developing rituals that will help you get started on a task (e.g., cleaning your house or office before you begin). You will find, too, that it's essential to make adjustments as you go.

Finally, we urge you to think not only about how you are doing at different stages in your research, but also about how your participants are doing. If you are not sure, ask them. Be sensitive to their experiences and to what they might be getting out of or taking from your research. Also, be appreciative. A very common notion in qualitative research is that of reciprocity, or giving something back to your participants. Sometimes this ends up being an actual gift, but reciprocity can also be something intangible; for example, your participants may learn something useful from your research or benefit in some way simply from being asked to reflect on and respond to your questions. Qualitative research usually entails give and take and, ideally, it is something that is beneficial both to you and to your participants.

APPENDIX A: STUDENT READING SURVEY

Reading Survey
English 9
2002–03

Name _____

Home Phone Number _____

Parents' Names _____

Parents' Work Phone Numbers _____   _____

Your Home School at PCEP _____

Middle School You Attended _____

Language Arts Teacher Last Year _____

How do you go about choosing things to read? Check all that apply.

_____ Recognize author's name

_____ Title grabbed me

_____ Cover is appealing

_____ Blurb on back is appealing

_____ Read a little part of it

_____ Recommended by a friend

_____ Recommended by a family member

_____ Recommended by a teacher

_____ Recommended by a librarian

_____ Interest in subject matter

_____ Interest in genre

_____ On bestseller list

_____ On a list of recommended books

_____ Award winner

_____ Read a book review

_____ Heard about it on the radio

_____ Heard about it on TV

_____ Read about it in a magazine

_____ Saw the movie

_____ Saw it on the shelf (in classroom or library)

_____ Bought it in a bookstore

_____ Received it as a gift

_____ It was in my house

_____ Part of a series

_____ Length

_____ Difficulty level

_____ Other _____

* Put a star next to one item on the above list that influences your choice of books the most.

What purposes can you think of for reading? (Reasons to read) List as many as you can think of.

List the titles and/or authors of any books you would consider to be your favorites.

List the types of books you like by category.

List any other things you like to read besides books.

Best thing about reading:

Worst thing about reading:

APPENDIX B: WRITING ACROSS THE CURRICULUM SURVEY

1) Please use the grid below to list the three undergraduate courses you teach most often. Also indicate in the appropriate boxes how often you have taught each course in the last 4 years, the kind of course, approximate enrollment, and if writing was required.

Course name and number	Number of times you have taught the class in the last four years	Kind of course					Approximate enrollment
		Lecture	Recitation	Lab	Online	Other	
1*							
2*							
3*							

*In the rest of this survey, please refer to your three courses using these numbers.

2) Please review the following competencies and select the three that you believe are most essential for students to have before registering for each of the courses listed above. Rank your selections one through three to indicate their importance.

Competencies	Course 1	Course 2	Course 3
Introductory knowledge in content area			
Critical reading/thinking skills			
Artistic skills (drawing, painting, musical, visual)			
Library/research skills			
Oral communication skills			
Interviewing skills			
Written communication skills			
Typing or word processing skills			
Internet Skills			
Other specialized computer skills			

3) Please indicate how important you believe writing is in all of your undergraduate courses.

	Not important				Very important
100/200 level courses	0	1	2	3	4
300/400 level courses	0	1	2	3	4

4) Please indicate your perception of the importance that members of your department attach to writing in undergraduate courses.

Not important				Very important
0	1	2	3	4

Comments:

5) Please indicate the importance you believe members of your department should place on writing in undergraduate courses

Not important				Very important
0	1	2	3	4

Comments:

6) For each of the three classes you indicated above, please check all the instructional strategies you use. Write most in the appropriate box to indicate the instructional strategy you use most frequently in each class.

	Course 1	Course 2	Course 3
Lecture			
In-class writing			
Small group discussion/work			
Computer work			
Videos/films			
Class discussion			
Lab discussion			
Presentations			
Performances			
Guest speakers			
Computer conferencing			
Other:			
Other:			
Other:			

7) For each of the three courses you listed above, rank the following categories for their importance in influencing your judgment of the final drafts of your student writing. Rank your selections numerically to indicate their importance.

	Course 1	Course 2	Course 3
Demonstration of critical, logical, analytical thinking			
Content (choice/definition of topic; application of class terms, concepts, themes; accuracy; use of examples or evidence)			
Depth and originality of analysis (substance; depth; creativity & originality of approach)			
Organization (general coherence)			
Demonstration of student growth, self-awareness			
Presentation (format; neatness; on time; style; appropriateness and accuracy of citation style)			
Mechanics or grammar			
Other:*			

*If applicable, please write in a category and include it in your ranking.

8) In each of the three classes you indicated above, if you assigned writing, please describe all of the writing assignments you gave. Please also check the type(s) of feedback students received for each assignment, as well as approximately how many total pages of writing you required.

Course 1 Assignments: (Some examples of assignments include research papers, lab reports, unit plans, outlines, creative writing projects, interviews, book reviews, reviews of a performance or work of art, web caucus, short or long answer essay responses on exams.)	Did students receive			Approximate number of pages: (Include in your estimate writing to help the generation of ideas, rough drafts, final drafts, and any work leading up to the final draft.)
	Instructor comments	Peer comments	Grade	

Course 2 Assignments: (Please see examples above.)	Did students receive			Approximate number of pages (Please see above.)
	Instructor comments	Peer comments	Grade	

Course 3 Assignments: (Please see examples above.)	Did students receive			Approximate number of pages (Please see above.)
	Instructor comments	Peer comments	Grade	
_____				_____
_____				_____
_____				_____

9) If you do not use writing in the three courses you teach most frequently, please indicate why.

10) Department: _____

11) Please check job title:

____ Professor ____ Associate Professor ____ Assistant Professor

____ Lecturer ____ Instructor ____ Other: _____

12) Have you participated in a Writing Across the Curriculum summer workshop?
Yes ____ No ____

13) (OPTIONAL) Name: _____

14) Would you be willing to participate in a Writing Across the Curriculum focus group? Yes ____ No ____

15) (OPTIONAL) E-mail address: _____

6

What Do I Do With the Information I Collect? Analyzing Data

By this point in your research, you have formulated a well-focused research question and presumably completed your literature review. You have also started gathering data by using at least one of the research tools we discussed in the previous two chapters. In other words, you have begun gathering information that addresses your question from a variety of sources. As a result, you are well on your way to accumulating the pile of transcriptions, observation notes, and/or artifacts and research notes that are at the center of the next step in this process: analyzing data. The task of sitting down to look at all of this material can seem daunting. You might ask yourself: How can I make sense of all the material I've gathered? How will I ever have the time or energy to look at it all? Do I even have enough material? Where will all the reading I did for my literature review fit in? Will others agree with or even find my analysis useful?

Although the task of analyzing data can seem overwhelming, we have found that it can also be one of the most satisfying parts of the research process. During this stage, you have the opportunity to look carefully and systematically at all the data you have gathered and you can begin making sense of it, especially in the context of the questions that gave rise to your research. What we hope to do in this chapter is provide you with some productive tools that will help you make sense of your research findings, both for yourself and for others.

GENERAL ISSUES IN ANALYZING DATA

The following general issues are important when analyzing data:

- Knowing when to start analyzing
- Learning to look deeply
- Getting organized with materials
- Finding time and space

Knowing When to Start Analyzing

Eventually, it is time to stop gathering information and start focusing on analyzing it. But when is the right time to do this? How do you know when you have "enough" material to start? Sometimes this is determined by external factors: the deadline for your thesis is fast approaching; the deadline for an article or conference paper you are doing is coming up; or the time that you had available to work with your participants is coming to an end (e.g., if you are a teacher researcher and the school year is ending). You might also have set deadlines for yourself that you are determined to keep, although we caution you here to be flexible. In any of these cases, you will hopefully know that it's time to move on from gathering information to analyzing it because you will have amassed enough information to begin understanding the questions you have raised.

Most researchers do at least some data analysis, regardless of how preliminary it is, as they gather information. By preliminary analysis, we mean the casual and incidental summaries and interpretations you may do along the way. For example, you might look over a completed interview transcript and begin noting issues that seem to be coming up in the interview. Or you might reread your observation log every few weeks and write summaries of what you are learning. Many researchers believe that this kind of "analyzing as you go" is vital: Reflecting thoughtfully about what you are discovering as you are discovering it allows you to formulate new questions for research participants, create a new focus for your observations, and/or consider aspects of your research through a new lens. Other researchers, however, believe that this kind of ongoing reflection has the potential to create problems: They would argue, for example, that trying to make sense of your findings before you collect all of your data may make you more inclined to see some answers and not others, or may lead you to view your findings too narrowly.

Our stance is balance. Doing some analysis as you go can not only help you begin focusing in a positive way on what you are seeing, it can also help you stay focused and be more productive as you continue doing your research. The trick is to make sure that any preliminary analysis you do is contingent and not cast in stone. Analysis is a way of looking and re-looking that, done throughout the research process, will continue to lead you to new ideas rather than to premature conclusions.

In the sidebar for this section, there are several strategies you can use to carry out the preliminary and ongoing analysis we are suggesting. If, after trying some of these strategies, you still find yourself asking questions to which you can't find answers, then you

Sidebar 1: *Strategies for Carrying Out Preliminary and Ongoing Data Analysis*

Field Note Summaries: Once every 2 weeks or so, it can be useful to reread the field notes you have taken in your research and write a summary of them, one that both makes sense of what you have noticed so far and also raises questions. Taking time periodically to reflect in this way can help you narrow or expand the focus of your research. As an example, in a study Cathy did in which she shadowed a 10th-grade student for a school year, looking to see how the student approached writing in her various classes, she one day wrote the following field note summary:

> Her rules for writing are not so much by subject area but by type and by teacher perception of type: most teachers tell the rules. Not by traditional modes of development but by final form: report, essay, summary, evaluation—defined generally in terms of parts and lengths rather than content.

Research Memos: Although they can take a variety of forms, research memos require the researcher to consider what she has learned so far from her sources and summarize that information briefly. Usually intended for an audience beyond the researcher (e.g., your advisor, an editor, a colleague), research memos allow the researcher, first, to try to make sense of what might be disparate findings and, second, to try to represent those findings to another person who might offer new insights on the research. The casual and questioning nature of these is just right for preliminary analysis. As an example, our student, Julie Caldwell, who was in the midst of her research on the vocabulary of her middle school students, wrote this memo to Cathy:

> Dear Cathy: This week was great in that I had interviews to work with and time to reflect.
>
> I am not sure what I'm driving at in the interviews, and I need to interview more students. I have had three interviews with Isabel, and the benefit is that she is really opening up and becoming more forthcoming. So that is good. …
>
> I'm really wondering where my data collection is leading, and I truly look forward to glimpsing trends. So far I haven't come around to seeing any, but I think it's because I'm not ready. I want to collect a few more interviews. Also, the students I've chosen to focus on are so different.
>
> As I peruse Isabel's interviews, I see that she is almost afraid to admit that she doesn't know words—she only wants to use words she thinks she knows—so I see this process, this sort of screen through which vocabulary passes into her radar. Then she will recognize it. She is very choosy about the style of words—diction really, I guess. Some words she "likes," others she doesn't. Hmm. That is so interesting to me. Shall I do a case study of Isabel?

Graphic Depictions: Sometimes researchers use sketching as a way to think about the information they have gathered. Representing symbolically what you know so far, what you still want to know, and how you will get there can help you see the big picture of your research and where to go next.

Prompt 1:   *Doing Some Preliminary Analysis.*

If you have already begun gathering data for your research, use one of the strategies described in the sidebar for this section to do some preliminary analysis of that data. Share your analysis with your classmates to get their feedback.

may need to do more data gathering. Keep in mind that the line between these two stages of research is a blurred one: Most researchers move back and forth between them.

Even if you have amassed a good deal of information from your research and have analyzed it all along, you may still have a hard time committing to and beginning the analysis stage of the research process. You may think, as we often do, "but there's more I could do: one more interview, one more observation, one more analysis of a document." If you find yourself hesitant to move on, keep in mind, first, that your ability to obtain more information is rarely cut off. You can still contact a participant one more time or review one more document if you need to. Also, even if you make them flexible, it can be very helpful to set some deadlines for yourself (e.g., saying that by December 1, you will have completed your interviews and observations and started your analysis of them). We also believe that it is much easier to begin your data analysis if you have a handle on some of the tools and strategies that will help you. We address these tools and strategies in this chapter.

Learning to Look Deeply

We liken the process of what you do with all of the information you collect to what the protagonists do in the mysteries we read. All good mysteries, of course, require the detectives both to make sense of the information presented to them and to move beyond the obvious suspect or motive. It is only by digging beneath the surface, trying to put the clues together in a different and unusual way, that the mystery gets solved.

That is how we view the analysis stage of the research process. Part of your job as researcher is to sift through all of the material you've gathered, trying to find the themes that emerge from the multiple research strategies you used and then using those themes, along with other evidence you find, to answer your research questions. This is the moment to look both *within* each of the research strategies you used and *across* them (e.g., to analyze both what an individual interview tells you as well as what all of your interviews tell you). This is when you begin connecting the outcomes of all the strategies you used, including the literature review. It is when you ask, "What have I discovered? What does the evidence I've gathered from all of these sources indicate?" A large part of your job at this point, then, is to begin looking for commonalities—how your multiple ways of looking into your questions may point to a common answer or answers. However, another part of your job at this point is to look beyond the commonalities and to notice the contradictions that may occur. This looking beyond commonalities is generally what will provide you with the kinds of complex understandings you need in order to answer your original questions even more fully.

But how do you know if there's a contradiction? Contradictions inevitably occur when we research real situations with real people. Seldom are such situations totally consistent. For example, a student you interview may tell you one day that revision is the most helpful part of the writing process; however, a week later he may complain that revision is stupid and tell you that he never even does it. A business memo that you analyze may refer to a particular set of office policies that management claims are always followed, but the individuals who work in that office demonstrate no understanding of those policies. Part of our jobs as researchers in these situations is to dig deeper, to try to find out why the contradictions exist. Did the student revisers change their minds about revision after getting a paper back from their teacher that was covered in red ink? Was the business memo never actually circulated to the office workers? Asking additional questions or observing the situation more closely can help you think deeply about what is really going on and why the contradiction might exist.

This kind of analysis is not always an easy task for at least two reasons. First, it's much easier and quicker to look through the evidence you have collected with a particular answer in mind and to focus primarily on the data that supports that answer. It is satisfying to find supporting information and to feel as if you have discovered a "truth." The other challenge relates to the fact that we all bring multiple perspectives, backgrounds, and experiences to our work that inevitably influence how we look at our data and what we see in it. In short, as qualitative researchers, we have both the advantage and the disadvantage of seeing the evidence we gather through our own lenses.[1] Therefore, what we see may not be what others see. We might then ask, "Is our answer or approach the right one?" If our findings can be viewed in multiple ways, how can we ever know that what we're seeing in them has any real meaning?

As long-time researchers, we continue to struggle with these issues. We worry that we have not dug deeply enough or that our interpretations of our findings might be idiosyncratic. We worry that our experiences and perspectives may limit our ability to look at our data with fresh eyes and open minds. We talk more about these issues in the lenses section of this chapter, but for now we just want to acknowledge that feeling some insecurity and even resistance at this stage is very normal. The most successful researchers are those who learn to accept and even revel

---

[1]We want to argue that although this criticism of personal bias in research is most often directed at qualitative researchers, we believe *all* researchers bring bias to their work. Our view is that researchers should always be attuned to the biases that they bring to their work—whether those biases are part of the questions they ask, the participants they include, the methodologies they employ, or the approaches to analysis they take. Researchers must also bring that awareness to the forefront of what they do. For more on this issue, see a later section in this chapter on lenses and triangulation.

in the uncertainty, through what Hubbard and Power (1999) call the "murkiness of data analysis": "The murkiness of data analysis is what scares any researcher. If it doesn't spook you at least a little, you're not opening yourself up enough to the new learning that can come during analysis. If the analysis seems very easy, you've probably only found what you already knew before the project began" (p. 117).

It may also help to remember that analysis, like every other part of research, is a process. There is no single correct meaning to be found in your data. By looking and re-looking at the information you have gathered, you will find that consistent patterns begin to emerge. This chapter offers you multiple tools for finding those patterns and for feeling more confident in looking deeply at the material that has been gathered.

Getting Organized With Materials

During data analysis, all of the ways in which you have organized your research up to this point will serve you well. Analyzing data is, in many respects, like putting together a puzzle. Some put puzzles together by randomly searching for pieces, but experience usually teaches us to adopt a more strategic approach; for example, to put all of the pieces that form edges in one pile and all of the pieces of similar color in another pile. Whereas serendipity plays a role—we occasionally just reach in and find a piece that fits perfectly—most of the time we need to look carefully to discover an underlying pattern.

Getting organized for data analysis means not only having the materials you need easily accessible (interview transcripts, observation notes, artifact analyses), it also means having time and space to immerse yourself in the data. Putting together the pieces of your research findings is like finding the pattern of a puzzle: It requires both time and careful attention on your part.

In order to make data analysis less overwhelming, it helps to do some organizing as you are gathering your data. If you have not done so, you certainly want to before you begin looking at your data. Different researchers have different strategies for organizing and keeping track of data. Next, we present a few strategies that we have found useful.

The first involves using the computer to get organized. These days, many researchers find it useful to put all of their materials onto the computer, creating separate files and folders, for example, for different kinds of data. Researchers can scan or retype observations notes into a file, or even use a laptop to make the notes in the first place. A key advantage to having your materials on the computer is that many of the analysis tools you use might themselves be on the computer, either as distinct software programs or as tools or functions within other programs that help you accomplish tasks like sorting through and categorizing information.

A second strategy entails organizing data using file folders. If you prefer to work with hard copies, one easy way to organize your data is to use different colored file folders depending on the source of the data. For example, you might use red folders for interview transcripts, with a separate red folder for each participant. You could use green folders then for observations, and so on. You may also choose to arrange your materials chronologically, alphabetically, by participant or setting, and so forth. One advantage to using file folders is that they allow you to quickly access documents so you can review and compare them easily. Of course, you can also do similar kinds of reviews and comparisons with a computer.

A third strategy entails organizing by notebooks. Some researchers use research logs or notebooks that they have written in by hand or kept on the computer. They use them for observation notes as well as for notes from interviews. The ideal here is to use different notebooks or files for different kinds of research. If you put everything into one notebook or file, then one way to organize the material is to place different colored and different sized Post-it® notes on particular pages or sections to indicate various kinds of information. You can place the Post-its® on the beginning page of each new type of material (e.g., red for interviews, green for observation notes, etc.). If you have kept electronic notebooks, you can easily divide them into separate files, label the parts for easy searching, and so on.

As we talked to a number of researchers about how they organize their data, what we discovered is that what works well for one researcher is not necessarily successful for another. What is most important is to create a system that works for you. The sidebar for this section presents several researchers' strategies for organizing their data. Note the variations.

Finding Time and Space

Another important issue in data analysis relates to the context in which you will carry out the analysis—in other words, the space where you will do your work and the time you will be able to devote to it. Analyzing data in the manner discussed here requires that you immerse yourself in the information that has been collected, that you go as deeply as you can into the information, and that you take the time that is needed to do this.

This is not the kind of work that can be done sporadically or in limited chunks of time, such as an afternoon here or there (although it's true that something valuable might occur to you in those moments when you're in the shower or walking the dog). Immersion generally requires large blocks of regularly scheduled time (ideally on a daily basis) so that you can lose yourself in your data. The reality of most of our lives, however, often precludes this ideal. Very few of us can put our lives on hold completely to do our research. The best advice we can offer is that you think carefully about how to

Sidebar 2: *Researchers' Strategies for Organizing Data*

Robert Yagelski, SUNY Albany: In the last few studies I've done, I've used Microsoft Excel as a way to organize notes, some data, and procedures. So in the case of one study that required taking field notes in a class, I would write the notes out in longhand during the class, then that evening rewrite them in MS Excel, using the program to organize the notes into the broad categories that my co-researcher and I were working with. In a way, I was doing some low-level data analysis as I rewrote those field notes.

Bob Fecho, University of Georgia: Most of my research is on my own practice and it's qualitative in nature. More and more, I keep both hard copies and e-copies of my data. Having the e-copies helps me later when I'm writing up my work or excerpts for close looking. I tend to keep the hard copies in milk-crate-like containers (I've found ones that allow for folders to stand up easily). In such a container, I can keep everything in files and/or binders and even keep any audio or videotapes. I've been using email quite a bit as a data source these days, so I keep mailboxes either by project name or by the names of participants in the project. I also keep a hard copy binder of every email usually in chronological order. In general, how I label any folder has much to do with the intent of the project.

Rebecca Sipe, Eastern Michigan University: I'm a notebook and box organizer. All the quotes, comments, and excerpts from the spelling research, for example, made their way into category sections of a notebook. Other pieces are categorized in file folders in plastic boxes that line the sides of my office. When I'm writing I can get to things quickly and easily that way. It looks like a real mess, but there is organization.

Todd Destigter, University of Illinois at Chicago: I … use a bunch of files in which I collect stuff that I'm thinking about. Once I begin to see the concepts "rise from the data" (i.e., once I start identifying … conceptual categories), I actually use different colored highlighters to flag specific stuff that would illustrate and help me explore/explain the bigger concept. For instance, right now I'm looking at a number of things at Latino Youth, among them how the few black kids feel about being at a traditionally Latino school. My notes and artifacts that deal with that are highlighted in light blue. The stuff that is about why this one particular girl I'm working with doesn't come to school is in yellow, etc. Every two or three weeks, I print out my notes and musings, punch binder holes in them, and put them in a notebook, where I also highlight them with the color codes.

carve out the time you need to do your analysis. Think about breaks in your schedule. Is it possible to work for an hour each morning before your work day begins? Or, how about taking an hour in the evening? Can you set aside one night a week? It also helps if you can set a regular time for your work and establish a routine for it.

Prompt 2: *Identifying Your Organizational Strengths and Weaknesses.*

Think about your own organizational strengths and challenges. How have you organized your academic or personal work in the past? What have you liked about how you organize your work? What problems have arisen in your organizational strategies? What changes would you like to make? Compare your ideas with the ideas of others in the class. What have they done so far to organize the information they have collected?

You should also think about the physical space in which you will do your work. Many researchers like to spread out their materials and have them readily accessible. This can be done electronically as well, but many researchers still like to have the physical materials spread out in front of them. Much like working on a jigsaw puzzle, we like to have all of the pieces handy so we can search through them, looking for missing pieces by hand, physically moving them around so we can visualize patterns and relationships, and moving back and forth among documents. All of these physical activities help us see how the various ideas, quotes, and sections fit together. Because most of us analyze data recursively, having the data physically arrayed so that it's easy to manipulate and manage is essential.

What you use for your physical space will depend on your own context and circumstances. You might set up a card table in a corner of a room, use one half or all of the dining room table, or use the floor in your family room or bedroom. Find some space that is both comfortable and workable for you and that you can take over for a period of time so that you don't need to keep cleaning up your materials. The ideal situation is to have a space that you can work in whenever it's convenient, someplace where you can easily come and go.

Prompt 3: *Getting Organized for Analyzing Your Data.*

First, create a calendar for your analysis work. Put in due dates and deadlines, add any additional tasks you need to complete (e.g., more reading, interviews, surveys), note times when you might be very available or not available at all to do your work (e.g., vacations). Next, create a weekly work schedule that indicates days of the week and times that you can devote to your work. Indicate what you will do during those times (e.g., Saturday morning—reread transcripts from initial interviews). Finally, develop a plan for organizing your research materials for analysis (e.g., will you create folders for each research method—interviews, artifacts—or for each participant). Also, make a list of the supplies you will need (e.g., Post-it® notes, file folders, legal pads, different colored pens, or perhaps a specific data analysis software program). Keep in mind that your calendar and schedule will likely change, so be prepared to be flexible.

TOOLS FOR ANALYZING DATA

The following tools are useful for analyzing data:

- Reading/Immersion
- General thematizing
- Coding, indexing, and writing research memos

Reading/Immersion

Now that you've carved out time, created space, and organized all of your materials, it is time to determine what to actually do to make sense of your data. One thing you need to do early on is to read through everything that has been gathered. There are various ways to do this. Initially, for example, you may read through all of your data just to get a sense of the whole and to get a general understanding of what your participants told you or what your observations revealed. At some point, however, you need to start reading more closely, paying attention to the particular meanings contained in your data. This kind of reading usually does not proceed linearly but, rather, recursively. You begin reading for specific information, seeking answers to the questions you have raised, looking for connections both within and across research documents, and looking for nuances (e.g., what was said versus what wasn't said or mentioned). In short, the reading of data usually is done on several levels. Researchers read broadly, searching for general themes and meanings and, closely, looking for subtle points and connections and going beyond what is on the surface. Such close reading requires enough concentration so you can make connections and recall earlier ideas. Most researchers also take notes as they read, constantly adding to and modifying their notes because reading is a cumulative activity: With each new reading, you bring with you the context of what you have already read.

We can't stress enough the value of reading the same material multiple times. We both have experienced that moment in our research when, in our third (or fourth or fifth) time over the same material, we suddenly saw the material in a new way. There is also value in spending time away from your data, time reflecting on and thinking about what all the various words and images might mean. It is important, therefore, both to immerse yourself in your data and to think deeply about it. Immersion and reflection are what enable researchers to make connections, even when those connections occur at times other than when you are actually doing your research (e.g., while you are walking the dog or cooking dinner). Thinking about your data when you are away from it can trigger a creative leap, moving you away from the obvious and allowing you to notice a previously unseen connection.

Prompt 4:   *Rereading.*

Return to a reading you did previously (e.g., for your literature review). Reread the piece, consciously slowing yourself down as you read. Take note of things you notice this time that you missed last time and look for connections between what you are reading and the data you have been collecting. Reflect on your activity: What was different in your reading this time and why? Also, how was it different and how did the meaning of the reading change for you?

Prompt 5:   *Reading Your Own Data.*

Select some of the interview transcripts, surveys, research log entries, or artifacts that you have collected for your research. Read through them several times. After the initial reading, write a summary of what you read. After each subsequent reading, record what you notice that you hadn't noticed the first time through. Discuss your experiences and findings with your classmates.

General Thematizing

The kind of purposeful reading described in the previous section leads to the essence of data analysis for qualitative researchers: creating a system to help make sense of the material you have collected. Reading, then, is an important initial step in data analysis. As a next step, qualitative researchers often begin looking for themes, a step that entails combing your data for ideas that keep cropping up. With this strategy, you look for general connections among and between the pieces of information that have been collected. As we explain what is entailed in finding themes, we will draw on the prompt for this section, using the responses you give to illustrate how to uncover themes in your own research.

Prompt 6:   *Finding Themes.*

Take 5 minutes to write about your earliest memory of learning to read. Describe as clearly as you can where you were, who was with you, what you and others did, and how you felt. Share your reflection with your classmates. As individuals share their reading memories, have someone record on the board the key points from each reflection. Look at what is written on the board and think about what you heard. What points of commonality exist across the stories? What themes seem to emerge from your collective experiences of early reading?

What you will probably notice in doing the prompt is that there are certain common themes in your early literacy stories. These commonalities might relate to *others who were involved in the literacy experiences*, to whether these *experiences were in or out of school*, or to whether the *experiences evoked positive or negative memories*. Common themes might also emerge around certain kinds of books (like Dr. Seuss) or certain television shows (like *Sesame Street*) that were a part of your youth. The themes you identify, whatever they are, can help you begin forming a picture about this concept of early literacy, a picture you could explore further by looking at other kinds of information: in-depth interviews with people in the class, observations of small children learning to read, discussions with children and parents about their reading, and so on. In other words, the general themes you identify can help you begin seeing what might be in your research and can help you situate yourself more deeply in the information you have gathered.

After identifying themes, many researchers move back into their research, looking to develop more specific categories from these themes. These categories are generally narrower than the themes. For example, working from the theme of *others involved in early literacy experiences of children*, you might identify several categories: parents, siblings, friends, and teachers. You can then use these categories to sift more deeply through your research, uncovering, for example, how these various others influenced the reading development of the participants in your research.

Cathy's project that focused on teacher outreach, described earlier, offers an example of this kind of analysis. The project looked at the question, "How can teachers be more effective in their outreach to parents so that parents can receive a more realistic picture of what happens in literacy pedagogy than they are currently getting from media reports?" Concerned that too much of parents' knowledge seemed to emerge from inaccurate or incomplete reporting of how reading and writing are taught, she began working with a group of teachers who were equally concerned about this issue. She interviewed these teachers about what they were doing and collected numerous artifacts from their classrooms, especially the handouts they sent home to parents. After pouring through the interview transcripts and artifacts, she started her analysis by listing all of the different means of outreach the teachers used (e.g., notes sent home with positive feedback about children, anthologies of student work with letters from the teacher, picnics at the start of the school year, books written by teachers about curriculum, etc.). After studying the lists she constructed and going back to her interview transcripts and notes, she began to recognize that the teachers seemed to have a staged approach to outreach, using different strategies at different stages. She eventually identified the following categories: strategies for community building, strategies for informing parents, and strategies for involving parents. Going back and forth between her research notes and these categories, she discovered that nearly all the strategies mentioned in interviews and demonstrated in artifacts fit neatly into one of these categories, which thus provided a way of organizing the material that was not previously apparent to Cathy or to the teachers.

What Cathy recognized, and what all researchers should recognize when they attempt to categorize their findings, is that, first, the process is recursive. The initial categories you develop will give you a way to look at the material, but re-immersing yourself in your findings may reveal additional or different categories. At that moment, your job is to ascertain what a new look into the material will tell you and how you can restructure your categories to better fit the information. It is not uncommon to go back and forth, changing and adapting categories as you come to understand your data more fully.

Also remember that not all of your material will fit neatly into your categories. Because qualitative research deals with the stories of real people and real situations, it is not an exact science. Seldom will there be entirely neat or exact fits. If there are, then that may be your first clue that you may be forcing a fit. As mentioned previously, these moments of discontinuity or contradiction are actually the job of this kind of research. When there is an exception, a piece of the puzzle that doesn't quite fit, qualitative researchers usually hone in on that piece, trying to figure it out. Exceptions often lead to the most important moments in research, the journey to figure out why the exception exists.

Coding, Indexing, and Writing Research Memos

Reading and looking for themes, and creating categories in a flexible and recursive manner, are important first steps in the analysis stage of research. Your next step as a researcher is to look through your data with a careful and critical eye to discover the evidence that supports the themes and categories you have identified. This sounds at first like a circular process: You look closely at the data to come up with themes, and then once they are identified, you go back to the data to discover evidence for them. The process is indeed circular, but we would distinguish the two steps in this way: You look initially at your interviews, observation notes, and discourse analyses to develop a general sense of the patterns that may be present. To begin discovering these patterns, you certainly have to read deeply, but your goal at this point is simply to discover general tendencies and connections. Once you move to the next stage, your job is twofold: First, you immerse yourself in the data even more fully (e.g., to consider how the details of the interviews, observations, and/or artifact analyses fit into the general themes and categories identified and, if necessary, to adjust and adapt those themes and categories). During this stage, it is vital to note specific moments, ideas, words, and phrases that support your themes and categories. Second, as you begin noting these moments, ideas, words, and phrases, you begin theorizing about the information you are uncovering; you begin considering "why": why these categories, why your participants responded as they did, why your observations revealed the information they did, and so on. Thus, while you search your notes for specific evidence to support the themes and categories you have named, you also start adding your own perspectives on possible questions and explanations.

Ruth Hubbard and Brenda Power (1999) distinguish between the notes you take as you research and the notes you develop as you analyze. They refer to the former as raw notes and the latter as cooked notes. "Raw notes," they explain, "are just what you've written, as quickly as possible, without any analysis. Cooked notes are the analysis of these raw materials" (p. 129). They continue:

> Cooking notes can ... be as simple as adding questions to them, to extend and expand your thinking about what you are seeing. ... Cooking with questions in mind extends your sight about what patterns are emerging. Questions to consider while cooking your notes might include, Why did I think this was important to write down? How does this connect with what I saw earlier in the day, week, year? Based upon what I'm seeing, what action should I take to change the curriculum or my research question? (p. 129)

The researchers we know use a number of methods to cook notes in order to deepen their analysis of the data they have collected. Some methods that are particularly useful are coding, indexing, and constructing research memos.

Coding

We define coding as a systematic way of indicating in your data the themes and categories that you have identified. In other words, with coding you mark all of the occurrences of the particular themes and categories that you have identified as important in your research, either through your early readings of your data or through your secondary research (e.g., reading of the literature and theory in your field). For example, for the early literacy prompt, you may have identified themes such as *influences of people on early literacy* and *influences of place on early literacy*. Perhaps some of the categories you identified for *influences of people* were teachers, parents, and peers. Under *influences of place*, you may have identified classrooms, bedrooms, and libraries as categories. Moving from this initial step of identifying themes and categories to marking these themes and categories through coding merely means returning to the data, reviewing it closely, and, while reviewing it, marking in some manner the various statements that relate to each of your themes and categories. Of course, it is also possible that you will find evidence while you are coding that does not relate to your themes and categories, so you will also want to make note of that evidence and consider adding new categories. It is important to remember that all aspects of data analysis are recursive, so you may end up revisiting your initial themes and categories. In other words, as you continue to re-immerse yourself in the information gathered, you may replace some of your categories with others that better describe the patterns you are seeing. Coding, then, can be a messy process because you will continue to move back and forth between the themes and categories you have identified and the specifics of your data.

When you code, you will also raise questions and do some initial theorizing about what the data means. In fact, your theorizing may be the driving force in the coding scheme you develop. For example, for the early literacy prompt, you may be influenced by theories of early brain development that suggest that the act of reading to children at a very young age is significant for a child's cognitive development. Holding this perspective might cause you to focus particular attention in your coding on the roles of parents and siblings in an individual's early literacy experience.

We also want to offer a side note with respect to coding. To reduce the messiness of coding and to keep yourself organized, make sure you have copies of the data you are analyzing. Because this step involves actually marking the documents that have been gathered, either electronically or by hand, and because you may very well end up changing your categories and/or how you are coding as you go, always have a clean set of originals on hand.

Ways to Code

At the simplest level, coding involves looking through the data and marking themes and categories in the margins. The marks indicate how particular words, phrases, and even segments of the data support the theme. The example in this section is taken from a response to the early literacy prompt (see Box 6–1). Our markings of it demonstrate how a researcher might code the occurrences of words and phrases that tie to the themes of influence of place and influence of people.

The first paragraph focuses on the library and conveys a clear sense of the influence of place. In her coding, the researcher begins to explore why this place is important to the respondent as an early reading memory: because it was comfortable with "chairs, pillows, comfortable rugs"; because it was "cool and inviting." In the next paragraph, however, a contrast appears. Although the researcher also marks this as example of the influence of place, she notes some differences as well. This part of the library was "all concrete and metal shelves … a little dusty." The less comfortable environment of this part of the library didn't seem to deter the respondent's desire to read: "sitting in a corner, a pile of books at my feet, and reading and reading." But the respondent does note that "this was the place for serious research" where the books were "hardback and nonfiction." Noting these statements helps the researcher ask the following question: "Nonfiction = less comfortable? I wonder if this association happened at other times?"

Box 6.1

One early memory I have of learning to read is going to the public library. I can remember **times when my mother took me** and times when I went with my class, but what I remember most is just being in that building. *I grew up in this tiny town, and so the library was even more impressive than it would have been had I grown up in a city.* The library was huge—with marble and wood everywhere. It had two children's rooms with wooden book cases filled with picture books and print books, and *it had space everywhere to hang out and read: chairs, pillows, comfortable rugs.* The ceilings must have been 20 feet high, so it always felt *cool and inviting.* **My mom used to go to meetings in the library, so I remember just being on my own in the children's room** and going through shelf after shelf of books. I felt so grown up when I got my first library card. It meant that I could take all these amazing books home!	Influence of place Influence of people Library mattered to her Comfortable place Fiction = comfort? Mom being there but also giving her time to be alone with the books
I also remember when I was a little older discovering the *stacks of the library.* You had to walk down these metal steps into the basement where there were rows and rows of books. I can remember sitting in a corner, a pile of books at my feet, and reading and reading. It was so different from the upstairs part of the library.	The influence of place is different from the previous paragraph. Non-fiction = less comfortable?
This part had *all concrete and metal shelves,* always seemed *a little dusty,* and the books were *hardback and nonfiction.* This was the place for *serious research,* I used to think.	I wonder if this association happened at other times?

Looking carefully at the words, fitting them to the themes and categories, and using all of this to ask additional questions is all a part of coding. A piece of data coded and organized in this way will help researchers look back on this response later and see how it connects to other responses they have coded.

As you do this kind of coding, you might use one of a number of techniques to make your job easier. Many researchers, for example, develop and use *abbreviations* (e.g., PL might stand for place; P might stand for people, etc.). Other techniques include *coding by highlighting*, where researchers use different colored pencils or highlighting and have a coding chart that indicates which colors relate to which themes. This kind of coding provides clear visual cues indicating which themes are occurring and with what frequency. The colors make it easy to summarize your findings by simply looking at the data. For example, if you want to ascertain how frequently a place is mentioned as an influence on early literacy, and if you have marked this category with a certain color highlighting, then it becomes very easy to flip through your data to note all of the instances you see. Color can also make it easy to identify quotes you might wish to use when you write up your findings.

Many researchers code using Post-it® notes or Post-it® flags and text markers. We know some researchers who stock several different sizes and colors of Post-its®, which they use throughout their research. During analysis, researchers can use different colored flags or the small rectangular notes to mark passages that connect to different themes and categories. Again, this approach can help you think more deeply about each category when you view your coding and analysis. Picking up an observation log or series of transcripts and flipping to all of the pages marked with a particular color can expedite your analysis.

Another option researchers have is to use a *coding chart,* as seen in see Table 6–1. These charts let you see how many times certain themes occur and how the themes move across the various kinds of data that have been collected. Generally, creating a coding chart involves these steps: naming the themes you are interested in looking at, naming the pieces of data to be examined, and noting the incidences of the themes in these pieces of data. It is also a good idea with this approach to leave room for questions and additional information. Returning to the early literacy prompt, a researcher, after looking at the responses of five people, could create a code sheet that looks like Table 6–1. The plus signs indicate evidence of the theme.

This kind of coding chart can let you see what themes and subthemes are most common and can help you think about why that's so. These charts can also help you raise questions about your data and categories. For example, Table 6–1 suggests the following questions: What is the role of night-time reading by parents? Do kids whose parents read to them like reading more as adults? Do they associate reading with comfort? What role do librarians play? Are they more prominent influences in situations in which parents are not influences on a child's reading? Such questions can lead you to review your data again as you seek a deeper understanding of these issues.

Whereas we are presenting these examples of coding as separate strategies, in reality many researchers use combinations of them. For example, it might be useful to code words or phrases first and then move on to creating a coding chart. Or you might find that highlighting works well for you, but that you want to add questions and ideas in the margins. The important point about coding is that it lets you mark the data in a strategic and systematic way so that you can keep returning to it as you dig more deeply into the materials you have gathered.

Coding also works best if you approach it flexibly, realizing that its purpose is to lead you to the next steps in the research process: making meaning of your research and then conveying that meaning to others. If your coding seems overly complex or difficult, then it may be a sign that your themes and categories don't truly represent what's in the data. If this happens, then you may want to return to your data and try

TABLE 6.1

Name	Influences of People: Parents	Influences of People: Teachers	Influences of People: Peers	Influences of Place: Bedroom	Influences of Place: Classroom	Influences of Place: Library
Michelle	+ (especially her mother, who bought her books every week)	+ (second-grade teacher, who suddenly helped her make sense of plot)		+ (found it "safe") nobody else in family read, so she could hide this "odd activity" from others		
Anthony		+	+ (competition to read the most books really got him inspired to read)		+ (couch in corner felt comforting)	
Michael	+ (mother took him to library every week) (I might need new category here for librarian's influence because he notes that the librarian helped him pick books)					+
Rachel	+ (mother read every night before bed)			+ (night time reading; comfort of bed)		+ (librarian took under wing) (I might need librarian category)
Carolyn	+ (father told stories every night before bed) (I wonder about the connection between story telling and reading!)					

to ascertain if there are other, more appropriate themes that cut across the information collected. Alternatively, if your coding seems to proceed too easily, then it might be a sign that you have gotten locked into certain categories early on and you may not be seeing all of the things your data has to offer. Again, it is probably a good idea to review your data multiple times to see what other information or issues emerge.

Prompt 7: *Practicing Coding.*

Choose one text from your research (interview transcript, log entry, writing sample). Read over the text searching for themes and categories. Make three copies and try out several of the strategies for coding. Did one strategy work better than another? Why? Did you find yourself using a combination of approaches?

Coding With Technology.   Coding, and qualitative data analysis more generally, are increasingly being supported by technology. There are a number of software programs available now that support qualitative research (searching in Google for "qualitative data analysis software" brings up several pages of hits). Quantitative researchers have long had statistical programs like SPSS available, but it has taken much longer for a standard to emerge for qualitative research (primarily, we believe, because it is so contextual and contingent). It is beyond the scope of this text to name and review all of these programs; however, we do encourage you to consider what might be available. For example, some universities purchase site licenses for particular programs in order to make them available free of charge to their students and faculty. Our university recently purchased site licenses for Nvivo 7, a fairly popular qualitative program. Other common applications that qualitative researchers use include NUD*IST, The Ethnograph, and ResearchWare. Some of these programs have free trial periods, so it may be useful to try them out first and see what they are capable of doing.

We caution that these programs are just that, programs, or tools, essentially. They can assist you with analyzing your data, but the expertise and skill that you bring to the analysis stage, and the theoretical perspectives you apply in analyzing your data, are still the key elements. You need, ultimately, to determine categories and themes and to develop and evaluate coding schemes that make sense for your work. You also need to interpret what you analyze and find. In other words, the intellectual tasks involved in analyzing qualitative data cannot be performed entirely by a computer. You also should be knowledgeable about what these programs can and cannot do and know about their advantages and limitations. Different programs have different strengths and weaknesses. Your job as researcher is to evaluate the tools to determine which might work best for you. Knowing as much as you can about these tools can help you make informed decisions that will enhance, and hopefully simplify (and perhaps speed up), your data analysis.

Indexing

Another approach to systematically reviewing your data is what some research-ers call indexing. Indexing is a way of keeping track of both the themes and the cat-egories you have identified in your data and where these themes and categories can be found. When you index, you produce a table of contents of sorts that records the themes and categories. Indexing can be particularly useful as a way to organize a lengthy observation journal or a series of interviews with a single individual. In-dexing is a tool that a researcher can use in conjunction with or independent of cod-ing. It sometimes follows coding and at other times is done by itself.

Hubbard and Power (1993) describe indexing in this way:

> The first step is to go back through your notes to date, reading through them and making notes in the margins. ... Then, on a separate sheet of paper, list the categories and themes that you have noted and each page on which these categories appear. After completing this index, jot down a few paragraphs about what you have noticed from these categories. (p. 74)

Dawn Putman, a student whose research centered around student reflection in writing, developed an index for her field notes, an excerpt of which, called "Dawn's Journal Entry Index," appears in the sidebar for this section. After reading her notes carefully and jotting down the themes she noticed in them, she eventually settled on seven categories that she believed warranted further study. She then went back through her notes and recorded the page numbers where examples of each theme could be found. For Dawn, this approach to categorizing her findings was impor-tant because it helped her see graphically the many examples she had found relat-ing to her various themes and categories. She was able both to support her belief that these themes were indeed weighty cues and to develop a system that would al-low her to return to those pages easily for further study.

Dawn used another kind of index to organize and make sense of the writing logs of individual students. After reading through one student's reflective writing log, she no-ticed that the entries weren't all of the same kind—that what she had initially called re-flection actually played out in numerous ways for this student. She began by noting the number of entries the student had written and then separated them into what she called *reflective* and *summative* entries (see the "Writer's Logs" entry in the sidebar for this section). Next, she began to think about the purposes for these entries and then named and defined categories for the various *purposes* she was beginning to recognize (see "Purposes" in the sidebar). Again, she counted the number of entries that served vari-ous purposes. Finally, she went back and noted the *key words* that the student actually used and inserted those into each category. This further elaboration helped Dawn un-derstand the student's purposes even more fully (see "Key Words Determining Pur-poses" in the sidebar). Dawn then went on to do this same process for all of the students on whom she was focusing for this study. Eventually, she was able to identify a pattern of word choice across her students' reflective writing; this knowledge helped her understand the topic of student reflection in new ways.

Sidebar 3: *Examples of Indexing*

Dawn's Journal Entry Index

Problems with student thought: 1, 2, 11, 14, 24, 35
Student motivation/investment: 1, 14–15, 52, 54, 51, 66–67
Question of whom reflection is for: 36, 35
Examples of reflective thinking: 2, 4, 14, 24–25, 27–28, 32–34, 49, 70, 72–74
Questions relating to reflection: 6, 9, 21, 25, 58
Questions relating to format of assignment: 29, 31, 47, 48
Problems with reflective assignments: 9, 10, 29, 44, 45, 50, 53–54, 55

Writer's Logs

Number of Entries: 80
    Reflective: 48
    Summative: 30

Purposes (note that some entries served two or more purposes):
    Evaluate—make judgments—22
    Compare—similarities/differences between tasks—4
    Plan/predict—determine what may happen—8
    Analyze—determine why something is how it is, draw conclusions—6
    Problem solve—determine alternatives—13

Key Words Determining Purposes:
    Evaluate: easier, good, love, don't like, I think, helped, hard, easy, helpful, harder, poor, interesting
    Compare: similar, than, compared to, easier [than]
    Plan/predict: I think, I will, going to, if I
    Analyze: realize, this is because, because, sometimes you have to, I realized, they seem
    Problem solve: I didn't know, I started … so, I might, I am, think about, it seems, I am having to, I can see how, that, I found, I began, I realized, I thought maybe

Writing Research Memos

As mentioned earlier, research memos can be useful analytic tools throughout the research process. They are designed to help you look through the data you have collected and write about what you are noticing. Research memos can also serve as useful tools as you are analyzing your data. A research memo is simply a report to yourself, generally written as a narrative, that summarizes the issues that have been identified so far and that explores how the data collected supports or raises questions about the themes you have named.

Whereas researchers often write memos to themselves as a way to come to terms with that they are seeing, it is also beneficial to write them to other audiences—a professor or a classmate, perhaps—as a way of learning what other people make of the analysis you have begun. In either case, the very act of summarizing your findings and attempting to explain why you have come to the tentative conclusions (or to the tentative questions) you have is an important strategy for making sense of your data.

We see the research memo functioning as a tool for digging more deeply into your data for at least two reasons. First, centering a research memo around a particular theme or category requires you to delve back into the actual words found in your observation logs, interview transcripts, and artifacts. As you reread these words, looking for the kinds of connections necessary to write the research memo, you will begin to uncover the patterns and explanations needed to answer your research question. A second reason to use the research memo for analyzing data is that it's a tool you can use throughout your research, thus giving you an opportunity to keep track of your interpretations of and perspectives on your data. In short, if you have kept your research memos over a series of weeks or months, then you may find that revisiting them at this point in the research process will help you keep track of your thinking and bring to light issues that might have gotten lost along the way.

Some variations of the research memo include the *one-page research memo*, in which you limit yourself to a single page, focusing on the very important issues in your research. Another type is the *dialogic research memo*, which can help you think through the evidence and arguments of two potentially important but contradictory themes. With these you can set up the memo on two facing pages, writing the rationale for one perspective and then the other. A third type is the *visual research memo*, which provides a graphic depiction of the work. These are useful for making sense of projects with complex interrelationships. They are especially useful with projects that have numerous themes connected in ways you have not yet ascertained. Finally, as with the other strategies discussed in this section, research memos are probably most valuable when used in combination with other strategies.

Prompt 8: ***Creating Research Memos.***

Look through several pieces of data that you have collected for your research. Try creating two kinds of memos from the information in those pieces of data: a narrative with an intended audience, a one-page memo, a dialogic memo, or a visual depiction. After you have written the memos, reflect on which seemed most useful to you. What could you capture in one and not the other?

VIEWING DATA THROUGH VARIOUS LENSES

In our minds, analyzing data is one of the most creative and exciting parts of the research process, an occasion for you to make sense of all the work done so far. Immersing yourself in the nuances of meaning and making creative leaps across findings can be a truly satisfying experience, an opportunity for you to begin to find answers to the questions you initially posed. After spending many hours developing a research question, doing a literature review, and gathering data through multiple means, you finally have the opportunity to see the big picture and to learn new information that might help you, and others, to understand a particular phenomenon or event more clearly.

We urge you, however, to approach this step with caution, ever aware of the lenses which, of necessity, inform your analysis. As all of us begin the search for connections and patterns, we run the risk of seeing our data through a specific lens that may shroud other viewpoints or that may cause the conclusions we reach to be a narrow representation of what is really in our data. Of course, as we have mentioned in previous chapters, we all approach the world with certain biases and predispositions. The analysis stage of research is not unique in this way. In short, no analysis is ever completely objective. However, as researchers, it is vital that you be aware of the predispositions you bring to this stage so that you can be vigilant about the effect those predispositions might have on your work. In other words, it is important to reflect continually on the predispositions and perspectives you bring and to acknowledge them in your work.

Earlier chapters suggested some prompts to help you think about your stances toward the world, both personal perspectives and more theoretical ones. As you reach this stage in your research, it may be useful to revisit those prompts as a means of thinking through how your various biases and perspectives may influence both what you see in your data and how you interpret what you see (see, in particular, prompts 9, identifying bias, and 12, identifying personal theories, in chap. 2).

One bias peculiar to this stage of research arises from your relationships with your research participants. You may find yourself, for example, giving more weight to the responses of certain individuals who were interviewed or observed because of your personal response to them: the woman who graciously invited you into her home and seemed to open up to you immediately, the work colleague with whom you had a complicated and sometimes hostile relationship, the teacher down the hall whose approach to kids runs counter to everything you believe about how children learn, the expert in the field whose work you admire and who complimented you on your articulate approach to your research question. Although we can't always help how we react to the people with whom we interact in our research, we can be aware of those reactions and make a concerted effort to approach their responses as objectively as possible. We have to ask constantly if we are valuing the responses of some participants more than others and why that's so. Sometimes it may help to write about why you seem to react in the ways you do to certain people, to try to separate out your response to someone's personality from your response to the information they are providing.

Prompt 9: *Reflecting on Your Biases in Your Relationships With Your Participants.*

Think about some of the people you interviewed or observed during your research. Are there certain people whose responses you valued more highly than those of others? Why do you think that was so? Are there people whose stances were very different from your own? How did you respond to the information they offered?

Being aware and self-conscious also applies to another frequent dilemma for researchers at this sage. Because as researchers we inevitably develop some preliminary interpretations as we collect artifacts, observe others, and conduct interviews, we may find ourselves formulating answers to our questions prematurely. When that occurs, we run the risk of approaching data analysis with answers already in mind. We are inclined, then, to see the data in certain ways before we even have a chance to really immerse ourselves in all of the information that has been gathered. We can, in those instances, easily miss some of the intriguing patterns and dissonances that are right in front of us. Again, whereas completely avoiding these moments of bias is not always easy, it is important to recognize that it is natural for them to occur. It is also important to take steps that will make us more aware of them either by talking with others or writing about and reflecting on them. The research memos that we discussed offer one opportunity for both reflecting on these moments ourselves and sharing them with others.

Another partial solution to these dilemmas entails triangulating your data in your research. *Triangulation* is a term that has many nuances, depending on the kind of research. In our view, triangulation means looking at an issue in your research from multiple perspectives. By multiple perspectives we mean, for example, that you consider a tentative conclusion about how middle school students spell by looking at the experiences of not just a few, but of several, students. If you are trying to determine how collaboration affects professional writers' writing style, it might mean that you look carefully not only at their own description of their writing, but also at your observation of their writing and their actual written products. In other words, triangulation implies looking at issues in your research in multiple ways to see if what you think you recognize through one source is also present as you look through others, whether those sources are other data collection tools or the experiences of other participants.

Another approach to triangulation is to ask others to look at your tentative findings so that you quite consciously draw on multiple lenses. For example, you may ask colleagues in your program, a professor, or a trusted friend to look through some of your materials to see if they see what you see. Many qualitative researchers even turn to their participants for feedback, to see whether their representations are "accurate," at least from the perspectives of the participants. This issue of participant feedback is important on many levels. For example, who better to give you their perspective than the individuals you interviewed or observed? However, it is also very complicated, which is discussed further in the next section. At this point, let us just say that the feedback you receive from participants can be a useful check on your own biases and one way to achieve triangulation.

Many of our students ask us if triangulation is achieved by simply looking at the data in three ways (three participants, three data sources, etc.). The answer is that it's not quite that simple. Triangulation means making your best effort to look at issues in your research from multiple perspectives. Further, the more perspectives you can bring, the better. Three is not a magic number. As a researcher, you should explore multiple perspectives on the issues you investigate as much as you can throughout your research.

Prompt 10: *Considering Others' Perspectives.*

Select one source of data that has intrigued you. You might see multiple ways of interpreting the information, you might be confused at contradictions in the material, or you might worry that your interpretation is leaving out some important piece of the puzzle. Share an unmarked copy of that source with several classmates. What interpretations do they bring? How are their interpretations different from yours? Can their explanations add anything to your own?

In addition to moments of personal bias, researchers also bring theoretical bias to data analysis. As has been mentioned many times throughout this book, we always bring theory to our research. When we analyze data, we inevitably read it through the eyes of our theoretical stances. Again, this is by no means bad. However, it is important that we be aware of our stances and cognizant of how they might interfere with how we view and interpret our data. For example, many feminist researchers in communication and writing are up front about their practice of looking at data with a woman's perspective and, in relation to classroom discourse, these researchers have discovered intriguing patterns that may not have emerged with a different perspective (e.g., see Finders, 1997; Gilligan, 1982). Likewise, researchers who are deeply concerned with justice might look at the same data in a very different light (e.g., see R. Bomer & K. Bomer, 2001). Researchers who work from particular perspectives are most successful when they identify for their readers the perspectives from which they come so that their theoretical lenses are not hidden.

ANALYZING DATA ETHICALLY

Whereas ethical issues should concern you throughout the research process, the ethical decisions you make during data analysis are serious and sometimes complicated. Especially if you are doing research that involves other people, complex issues about representation will inevitably arise—issues that have led researchers over the past decade to think hard about what research means and who has the right to conduct research. Essentially, the argument is this: Researchers who gain access to the words and lives of others have an ethical obligation to include those others in the process. Referred to sometimes as emancipatory, or participatory, research, this paradigm suggests that researchers should practice *research with* rather than *research on*, making sure that the conclusions they reach truly tell a story that takes into account the perspectives of those whose beliefs and lives they are attempting to represent. Unless research is participatory in this way, the argument continues, researchers reduce research participants to lab rats or guinea pigs, whose only role in the process is that of distanced subjects.

If we agree with this premise, then the implications for data analysis are huge. Who actually gets to tell the story of participants' experience: the researcher or the research participant? Whose themes, selection of quotes, arrangement of details, and biases and subjectivities get to count in the retelling? Many researchers have begun to argue that data analysis should be done in collaboration with research participants so that both the researcher and the participants can share ideas and come to some agreement.

However, the question at this stage of the research process (and sometimes in the writing stage as well) becomes how to do that kind of collaboration and how to achieve a co-construction. What are the obligations involved in that? And what are the mechanics of how to make it happen? Also, is it even always feasible or desirable? Again, a number of researchers have taken on these issues, and the suggestions they offer are important ones. Primarily, they suggest that we include research participants at various points along the way in the research process, especially during data analysis. Asking participants to read through our tentative conclusions, engaging them in conversations about our findings, or even asking them to write their own analysis of the data can help us create more participatory moments and, ultimately, a fuller interpretation of our findings. As Ann wrote, with Cole and Conefrey (1996), in "Constructing Voices in Writing Research: Developing Participatory Approaches to Situated Inquiry," certain strategies can help, such as asking research participants for clarification, asking participants to read the research searching for their own voice, modifying portions of the texts if the participants find them questionable or inaccurate, making certain to demonstrate the expertise of participants in their own lives and ideas, and stepping back from promoting oneself as "the" interpreter of knowledge (p. 147).

However, even with this self-awareness, dilemmas can arise when a researcher is confronted with the realities and complexities of negotiating meaning with participants and of understanding and taking into account multiple perspectives when analyzing data, some of which we might even take issue with. What happens, for example, when there is a conflict in the interpretations: Is the researcher understanding a particular idea one way because of the expertise she holds or because of the many interviews, observations, and readings she has done, and are the participants perhaps understanding the story differently because of their view of it? Katherine Borland (1991) talks about this dilemma in her important article, "'That's Not What I Said': Interpretive Conflict in Oral Narrative Research." Borland demonstrates the best of what researchers suggest in regard to including the words of others: Offer research participants a chance to see what we have written, what conclusions we are coming to, and how we are interpreting the information that they have offered us. However, what Borland's article also shows us is what can happen when a research participant—in this case her grandmother—is offered that opportunity and disagrees vehemently with the researcher's interpretation. Borland's grandmother sees the writing as a "definite and complete distortion, … " claiming that Borland "read into the story what you [she] wished to" (p. 70).

We wish we could offer you easy answers to the ethical dilemmas that arise from this kind of approach, but we cannot. Every qualitative researcher we know struggles with these same issues and comes to different conclusions. Borland ended up modifying some of the words, but her own interpretation prevailed. For Robert Yagelski (2001), a literacy researcher whose conclusions in a published article were questioned publicly by former students in the class he was describing (the students claimed he didn't do enough to include the voices of others), the answer lies in the concept of representation in the first place: "I know that neither their story nor mine is anything but a construction of the experience that inevitably misrepresents it. And to include more or different voices would only change that misrepresentation rather than make it 'more true'" (p. 657). For us, the best a researcher can do is to keep in mind the specific suggestions offered earlier: to ask research participants about the interpretation you have arrived at and about how their interpretations might intersect and differ. And, as we listen to their responses, we need to strive for a balance: We need to balance our own expertise with the expertise of those we have asked to tell us about their lives. We also need to listen attentively and with open minds to what our participants have told us, to question our interpretations constantly, and to realize that *truth* is a pretty elusive concept. In short, we need to acknowledge and appreciate that, at best, the answers to our research questions are carefully constructed representations. The way we construct and present those representations to various audiences is the subject addressed in chapter 7.

How Do I Present My Research? Writing Up Your Findings

As we move through the research process, we write continually. We write to think our way through our ideas, to formulate a research question, or to raise questions about our biases and beliefs. We write to record our findings as we read, conduct interviews, and carry out observations. We write to sort through our findings and make sense of our ideas as we analyze data and uncover themes and subthemes. Writing, in fact, is one of the constants of the research process, and up to now you have no doubt been doing all kinds of writing to develop your thoughts and ideas.

Writing is also an activity that presents researchers with numerous choices. In fact, writing can be characterized as a process that entails a succession of choices. This chapter addresses the choices you will make as you write up your research and presents concrete strategies to help you with those choices, emphasizing throughout that these choices are seldom binding. We also wish to emphasize that our presentation is by no means exhaustive. The goal is to help you start thinking about, planning for, and writing up your research in ways that will be most useful and beneficial to you, and to any audience you seek to address.

GENERAL ISSUES IN WRITING UP YOUR FINDINGS

The following general issues should be considered when writing up your findings:

- Shifting from writing to explore to writing to present ideas
- Taking stock of what you know already about writing
- Writing to interpret

Shifting From Writing to Explore to Writing to Present Ideas

Much of the writing you have been doing thus far has probably been more along the lines of *writing to learn*, that is, writing that is used to explore and discover, to ponder, to raise questions, and even to make sense of something—in this case your research question and findings. This type of writing is intended to be tentative and exploratory, a stopping place where you can pause and think through what you are learning. But, even if it is tentative, this kind of writing is integral to interpretation. When you write in this way, you are trying out an array of ideas and attempting to impose an order on those ideas as a strategy for making sense of them. For example, as you wrote various drafts of your question, as you wrote your research proposal, and as you started writing research memos and preliminary notes, you were at the very early stages of generating ideas and meaning through your writing. Usually, this early writing is intended only for yourself or, perhaps, your advisor. This early, less formal writing, hopefully, has already helped you move forward in your research and even led you to some initial interpretations of your findings.

At a certain point in the research process, however, your writing will begin to move from that less formal, exploratory writing to a more formal kind of writing that represents and communicates your interpretations—that is, from writing that has been a personal way to think through your ideas to writing that will present those ideas publicly. As such, your writing will now take on certain rhetorical features and motives. What, for example, is the *purpose* of your writing? (Are you trying to report on a situation? Make an argument for another approach? Offer advice to your coworkers?) What *kind of document* might you produce that will best mesh with that purpose? (Does your program have a set form, such as a thesis, or do you have flexibility in what you produce?) Who is the *audience* for your write-up? (Is it only for your advisor or committee, or will you try to publish it?) And, finally, how will you *represent yourself* as the researcher/writer of this document? (Will you present yourself as someone with a certain amount of expertise or as someone who is still learning?) All of these questions will affect the way you present your months of research.

Many decisions you will make about that presentation go back to your purposes for conducting the research in the first place. Although we talk about this later in the chapter, for now just think about your reasons for doing the research: Did you embark on this journey for personal or professional reasons? Did you hope to learn something that would help you in your own professional situation, or did you want to share knowledge with others? The answers to these questions will help you make decisions about the kind of document you want to produce (e.g., a scholarly piece, a more anecdotal piece, a report, a Web site, etc.) and the way in which you will go about producing it (e.g., the style and tone you will adopt, the approach you will take, the ideas you will emphasize, etc.).

Taking Stock of What You Know Already About Writing

As you move toward writing at this end stage of your research journey, keep in mind everything you already know as a professional in the field of writing. Sometimes our students—knowledgeable as they are about the nature of writing, the processes inherent in writing, and the best ways of teaching writing—forget these understandings when it comes to their own writing. Drawing on what you already do know about writing and applying that understanding to your own work can help you move through this stage with greater confidence and success. Consider, for example, what you know already about the nature of writing. As a student of writing, you know that writing is indeed a process, and that it takes time to draft, redraft, revise, and edit. Time, then, is a necessary consideration at this stage in the research process. You need to give yourself sufficient time to think and reflect and even to set aside your writing to obtain some distance and a fresh perspective on it.

It is also important to be aware of your own process as a writer, the ways you generally approach a writing task. Even though this project might be longer and more in depth than anything you have written in the past, the basic processes of writing you have developed over the years will remain the same. For example, some of us need feedback from others at all stages of the writing process—especially in the beginning stages. Cathy, for example, always talks through her initial ideas with a trusted colleague or with her husband. Just talking them through helps her crystallize her thoughts. Others find that freewriting their way through ideas is the best way to come up with a solid plan. The simple act of writing—with no constraints and no pressure to make perfect sense the first time through—helps these writers create meaning out of what may at first seem like disparate ideas. Still other successful writers use visual approaches to make sense of their ideas, finding the process of making maps, webs, or even outlines a useful means for finding their focus. And, of course, most writers have rituals in which they engage before writing. Some people need clean desks or studies when they write, some write best at certain times of the day, some have to clean the entire house before writing, and some need to answer all of their e-mail (which Ann always does). Whatever your process or ritual for getting started, accept and acknowledge its role in what and how you write.

Equally important to consider is how you approach the task of revising your writing. Most of us find that revision is a key component to producing a thoughtful piece of writing. Especially when you are immersed in a complex, often lengthy piece of writing, revision becomes an essential means of writing your way into a growing and developing understanding of your findings, as well as a way to continually refine your thinking. Revision in this sense speaks to the recursive nature of writing: Each revision takes you deeper into your finding—re-reading, re-thinking, and re-drafting help you to reach this depth of understanding.

Prompt 1: ***Who You Are as a Writer.***

Think for a moment about the last time you wrote a major paper or document. Try to picture where you were. Also try to remember your thought processes and the actual steps you went through in the writing. Respond to these questions:

- How did you get started with your writing? Did you take notes? Write an outline? Make a web? Talk with a friend? Freewrite? Just start writing?
- How did you revise? Did you share your writing with a friend or colleague to get their feedback? Did you revise as you wrote? Did you get a whole draft done before going back to revise? What did you revise for: Ideas? Sentence structure? Meaning? Organizational patterns?
- What parts of the process were successful for you? What parts didn't work so well?
- How and where did your best composing occur? At a computer? With paper and pen? In an office? At a coffee shop? In a kitchen? Sitting in bed?

Share your responses with others in the class. Do you have similar or different approaches to writing? Did you learn any strategies from your classmates that might be useful for you to try?

Writing to Interpret

Each draft you produce of a writing project should make it clear that writing is always an act of interpretation. As your write your way through various drafts, you will continually see and re-see the data you have gathered, and you will continually interpret and re-interpret the events you observed. Thus, how you write up your research—what you name as the themes and the important findings and how you situate and present those—may well be different from what any other researcher might do with the same material—and that is how it should be. At first, this might seem a little disconcerting and even counterintuitive: After all, shouldn't most research lead to an answer? Isn't that the point of research? Most qualitative researchers who understand the nature of the research they do know that this is not the point.

We can read and interpret research findings in multiple ways, just as we read novels, newspaper reports, and Web sites in different ways. Ethical researchers recognize this fact and make their claims, and qualifications, about what the research shows accordingly.

Anthropologist Margery Wolf would argue that even one researcher might interpret the same set of events differently, a stance she makes concrete in *A Thrice Told Tale* (1992). She interprets a similar set of events gathered from her research in three different ways, using different genres of representation in narrating the events: a short story, a set of anthropological fieldnotes, and a social science article. Her point, of course, is that interpretation is an inherent part of the research process and the rhetorical act of writing supports this notion of multiple interpretations. The events you choose to stress, the themes you choose to follow and highlight, and the genres you choose to set forth your understandings are integral choices in the research process—and, she would argue, always rhetorical.

Clifford Geertz makes a similar point in *Works and Lives: The Anthropologist as Author* (1988), focusing specifically on one part of the rhetorical construct: the nature of authorship in research. He suggests that we would do well to consider the researcher as an author: as a creator of texts who makes particular decisions about what to include and exclude, about what to emphasize and de-emphasize. He suggests that any anthropological reporting is indeed a kind of argument in which anthropologists have to, in part, "convince us ... not merely that they themselves have truly 'been there,' but ... that had we been there we should have seen what they saw, felt what they felt, concluded what they concluded" (p. 16). Thus, it is important to recall that as researchers we are also authors whose interpretations of events may indeed vary but who are always engaged in an act of interpretation. And our interpretations, although varied, are no less significant because of that variation.

So what does all of this mean for you as a researcher? First, it implies a certain level of responsibility on your part as you work hard to make sure your interpretation is supported by the information you have collected (we talk more about this later). If writing is an interpretive act, in other words, you need to be very sure that your interpretation emerges from and is supported by your data. Second, it suggests that because multiple interpretations are indeed possible, you might use writing as a way to sift through those possible meanings until you find the one that seems to be most supported by the data, that best meets the purpose and audiences for your writing, and that just "feels right"—that feels as if it works best to explain what you have learned. And, as we talked about in chapter 6, keep in mind the interpretive lenses that inform your stance on the world. As a feminist, as a Marxist, or as a believer in a particular approach to teaching or a particular approach to organizational strategies, you will most likely read your research through your particular way of seeing the world. Writing about those lenses in order to demonstrate your awareness of your stance, or even writing a section of your document through another interpretive lens (like Margery Wolf did), can be a useful means to evaluate and think more carefully about your approach.

The next section offers some suggestions for getting started with your writing and some tools that may help the writing process go smoothly. What is most important to remember, however, is this: As a professional in the field of writing, you are an experienced writer already. Whereas the scope of this project might be bigger than anything you have tackled before, all the strategies and tools that you know already are what will help you produce a final piece that you can be proud of and that others, if you are addressing others, will find useful.

TOOLS FOR WRITING UP YOUR RESEARCH

The following tools are useful when writing up your research:

- Taking a rhetorical approach
- Determining your focus or cut
- Being convincing/writing persuasively
- Obtaining and using feedback
- Letting your writing sit

Taking a Rhetorical Approach

Writing is, as we have stressed, a rhetorical act. Thus, before you immerse yourself in writing up your research, you should first consider your project from a rhetorical stance—focusing on its audience, purpose, and genre. This section offers a few ideas about each of these, and then asks you to consider them in terms of your own project.

Selecting Your Audience

We think it is important to begin with the question of audience. For some of you, the target audience might be your advisor or project committee. In our program, that's certainly one audience—and for good reasons (beyond just receiving a grade). It is vital to have an audience of knowledgeable professionals in the field who can read and respond to your research. This response will help you understand more fully what you have produced and how it contributes to the scholarly conversation of the field. Your advisor or committee can help you see your research in ways that expand its possibilities.

However, as we have emphasized throughout this book, we believe it is important to see your research as doing more than just satisfying a program requirement. We hope that as a professional in your field, you take on the mantle of research in your everyday work life. When that happens, your research becomes meaningful beyond just the academic requirements—in part because the audience for your research becomes even more genuine, as you try to reach those who might truly benefit from it.

Who are those who might benefit from your research? One significant person, of course, is you. Even beyond the exploratory and discovery writing already discussed, it is important to write up something that helps you understand better just what your research has shown you about your question—especially if the research is designed to help you change your own teaching or professional practices in some ways. But your audience, in most cases, will also expand beyond that. In our program, for example, we encourage students to write up their research for an audience that goes beyond themselves or their advisor: their colleagues in their own workplace, a supervisor, or colleagues around the country. We encourage them to start by thinking about whom their research might benefit or who might be interested in it. Writing for broader audiences can help you see the value of your research in different ways: The value becomes that of sharing with others what you have learned and/or adding to the growing body of knowledge in the field. Lawrence Stenhouse, long considered the "father of teacher research," saw this move to go public with research as a kind of ethical consideration. He believed teachers, for example, have an obligation to publish their findings, thus contributing to the "community of critical discourse" (Ruddock & Hopkins, 1985, p. 17) that exists in a field of study. We think that all professionals have the potential, and even the obligation, to contribute in that way.

The first step when you begin thinking about your audience is to consider who might benefit from or be interested in your research. This will help you identify your audience or potential audiences. Second is thinking about the needs of those audiences. What might be their questions about your research? How will your research affect their professional or personal lives? What background knowledge do they already have in the area, and what background might you need to provide? How might their situations and their interpretive lenses affect how they make sense of and respond to your writing? Although we realize that you won't be able to speak to and answer all the needs, questions, and contexts of everyone who might read your research, considering these questions before you begin writing can help you to start visualizing the kinds of information that will be necessary in your final write-up. It will also help you select an appropriate genre for your research, which we say more about later in this section.

One final note on audience: We have discovered over the years that we often write up our research differently for different audiences, using the same research findings but emphasizing different parts of those findings, or situating them differently in relation to the literature, in order to disseminate them to different groups. Ann, for example, has published her research on scientific discourse both in rhetoric journals and in more interdisciplinary journals in the social studies of science. For her publications in the latter, she made sure that she situated her work in a broader literature. Cathy, who regularly writes for audiences of both practicing K–12 teachers and English education scholars, has adjusted both her findings and the way she writes about those findings to meet the needs of these two groups.

As you think about the audience you want to reach, you may find that you have in mind several, very diverse, audiences, whose needs and questions might be quite different. In that case, you may select a "first" audience for your project with the intent of addressing other audiences in subsequent write-ups. Sometimes, too, it is simply easier to write for one of your audiences first and save other audiences, perhaps those that are less familiar and more difficult to address, for later. For example, one of our students conducted a rhetorical analysis of Nobel Prize lectures in chemistry for his culminating master's research. His first audience was his advisor and thesis committee and his initial document was his thesis, which contained a detailed analysis of the lectures and a detailed account of the science that led to the prize for the three chemists. Subsequently, this student began adapting his thesis for publication in a refereed journal in his field, a task that he found to be much different because of the expectations of the journal's audience—they wanted much less detail about the science and a more succinct and sharply focused rhetorical analysis. They also wanted to know what the work was contributing to the field.

Prompt 2: *Identifying Potential Audiences.*

Make a list of as many potential audiences for your research as you can think of.

Next, consider what you know about the backgrounds and needs of each audience as you start to fill in the following chart:

Audience	What might they already know about your research?	What might their questions be?

Determining Your Purpose

At this point, you might be asking yourself, "I thought I already determined my purpose when I articulated my question." While this certainly is true, it is also important to consider purpose again at this stage because you now need to consider your research findings in light of a larger goal: What can and should you do with the materials you have gathered and the answers you have begun to formulate. Your purpose, for example, might be to educate a particular audience about your research; it might be to suggest a fallacy in current practice; it might be to suggest a change in practice; or it might be to raise new questions about an issue.

And so, although we are listing it as a second step, determining the purpose of your research is really something you do in conjunction with determining the audience for it. Your audience and their expectations can sometimes help you determine your purpose ("What would this audience want to know about my research?"); however, (?) clarifying your purpose can also help you determine or better define your audience ("Who would want to know what I've discovered in my research?"). Thus, we recommend going back and forth between thinking about your audience and thinking about your purpose.

It might be helpful to hear what some of our students have seen as the purposes (and corresponding audiences) for their research:

- A high school teacher wanted to help other teachers in his building understand what writing across the curriculum could look like in practice.
- A PhD student in rhetoric and digital media wanted to share with other scholars in the field his findings about the epideictic functions of the Nobel lectures of scientists.
- A professional communicator and teacher of writing at a community college wanted to help students in technical and professional writing programs better understand the job market in those fields.
- A high school teacher wanted to help parents understand the choice reading practices she was using in the classroom, which were different from what any other teacher was using.
- A technical communicator wanted to help graduate students and colleagues learn more about international technical communication.
- A college composition instructor wanted to help other writing teachers develop a better understanding of how students actually learn about and begin addressing audiences.

For all of these researchers, the naming of their purpose added a narrower lens to the research they had done. For example, Jennifer Buehler, the high school teacher who wanted to help parents understand the reading practices she was using, began her research by simply asking the question, "What happens in my classroom when students are able to choose their own books for reading?" As her year progressed, she realized that she was spending a lot of time explaining to parents why she believed that this way of teaching was better for kids. At the same time, the data she was gathering strongly supported her approach. As she spent time the following summer sifting through her data, she realized that an important purpose of her research might be to help parents visualize her classroom: how many more books students read over the term, how they learn to select and respond to books, and how they feel about reading at the end of the year. Jennifer's purpose for her research became more refined as she began to think about how to represent her findings for a particular audience. She certainly could have had other purposes for that same research (e.g., helping other professionals in the field learn about the strategies she uses to help students read more or helping her own department create a new structure for their ninth-grade English classes). Her purpose—in this case determined in conjunction with her audience—set a specific direction for her writing.

As another example, our student, Erin Snoddy Moulton, started off her research wanting to obtain a better understanding for herself of international technical communication and translation practices. Her initial purpose was to expand her own professional knowledge so that she could potentially work in this area. As she carried out her research, she realized that her peers (other graduate students and professionals) could also benefit from the information she was finding. She refined her purpose, therefore, to include this broader audience. Instead of simply expanding her own knowledge of these issues, she became interested in helping other students become more knowledgeable about them. Like Jennifer, she could have also had other purposes for her research, such as contributing new knowledge to the field about globalization and translation practices. The purpose she selected shaped her writing as well as her choice of a forum (she ended up presenting her work in a Web site).

Prompt 3: *Identifying Purpose.*

Go back to your answers in prompt 2 and think about the purposes you have for your research and how those purposes might connect to specific audiences. Complete the following:

Audience	Their needs/questions ... ?	Your purpose in writing ... ?

Considering and Selecting a Genre

A third consideration when getting started in your writing is the genre you wish to use. Although as students you may be bound by the constraints of a thesis or academic paper that might be a requirement for your graduate program, at some point you will have the flexibility to expand the notion of genre beyond that. In our program, as mentioned earlier, we ask students to choose a genre for presenting their final project, and students have selected a wide variety, including:

- Articles for professional journals
- Handbooks for others in the field

- Conference presentations for local, state, or national conferences
- Web sites for parents, students, or other professionals
- Manuals (these have included more general manuals for parents or other teachers as well as technical manuals for end users of technical products)
- Curriculum documents for teachers or administrators
- Reports (again, for various kinds of audiences, such as funding agencies, administrators, peers)
- Proposals for funding agencies

What you may notice immediately when you look at this list is that certain genres fit particular research questions, particular audiences, and particular purposes better than others. To continue one of the earlier examples, when Jennifer decided to direct her research findings about her new curriculum ideas toward an audience of parents, she thought hard about the best way to get that information across to them. Ultimately, she selected the genre of a handbook and created a parent-friendly booklet that she gave to her students' parents the following year. Within that handbook, she was able to incorporate her research in multiple ways: In one section she described a typical day in her reading classroom, in another section she reported how her students used writing to demonstrate their understanding of their reading, and in another section she offered statistical evidence of the increase in student reading over the course of the semester.

Some of our students over the years have worried that the genre they select can't possibly represent all the data they have gathered and the new knowledge they have gained. We remember one student, for example, who really wanted to write an article for a particular professional journal, but was concerned that the brevity of the articles typically found in that journal (8–10 manuscript pages) would limit what she could say. As researchers and writers ourselves, we certainly recognize this dilemma—and would add that it can be very difficult to get enough distance from one's research at this point to be able to distill all of one's findings for a short article. One suggestion is for students to write a longer piece first and then consider writing in a more "compact" genre sometime in the future.

Choosing the right genre, then, also depends on the time you have available and your ability to synthesize and make sense of your findings in that time. Even as we write this sentence, we recognize the contradiction that exists between producing a thoughtful piece of writing in a particular genre and the reality of meeting a deadline for your graduate program. Again, here is where we suggest that you think about your research goals—one genre you produce from your research might fulfill the requirements of your graduate program, but producing another more thoughtful piece might become a long-term goal. Indeed, writing up your research in one genre can help you with writing in another. For example, both of us have experienced what is very common among researchers: Our first attempt at writing up research findings might be for a short conference paper in front of colleagues around the state or country; the experience of writing that and receiving feedback from peers might lead to our writing an article for a journal; eventually that article might be the beginning of a book. Each genre, then, serves a particular audience and purpose, but each also helps us make progress in thinking about our findings and their significance.

We believe it is very important to select a genre that will be useful to you in some way, which will serve you well as a professional, meet the needs of your designated audience, and satisfy your purpose. In order to do that, consider many possible genres before zeroing in on a certain one, keeping in mind that your findings can be used for multiple genres down the road.

Prompt 4: *Selecting a Genre.*

Look back at your answers for prompts 2 and 3. For each identified audience and purpose, consider a genre that might make sense. You can use the following:

Audience	Purpose?	Possible appropriate genres?

Look over the genres you've brainstormed. What do you know about them? Pick a specific genre and find four to six examples of it. What are the essential characteristics of the genre? What are its constraints? Do some of the examples seem better than others? Why? Share what you have learned about this genre with others in the class.

Positioning Yourself as an Author

One issue that often arises for our students is how to position themselves in relation to their research—in other words, the kind of self to project in their writing. Should you construct yourself, for example, as an expert in the field? Should you construct yourself as an expert in your research topic? Or, should you put yourself forth as a learner, a questioner? Should you write in the first person, making visible your role as researcher? Or, should you present your findings using a more distant third-person approach?

Again, the rhetorical nature of writing requires that you, as author, make conscious decisions about your position—decisions that, by the way, are not unique to new researchers. To illustrate what we mean, look, as an example, at the following passages from various research studies, passages that lay out for the reader the purpose of the study. (All of these were published in the journal *English Education,* a journal that Cathy co-edited for 5 years.) As you read them, think about the author(s) in the texts and how they represent themselves—both in terms of their research and in terms of the field at large.

Janet Swanson. In this essay, I argue that the teachers and teacher educators who participated in WFYL [Write for Your Life, an online teacher network] created an on-line "transformative teacher network"—that is, their on-line dialogue not only wove a web of connections among individual educators (the network), it also exhibited the characteristics of a "healthy network culture" that allowed idiosyncratic events to become occasions for authentic professional development, resulting in transformations in teachers' beliefs and practices and in student learning. By analyzing the dialogic "web" these teachers spun across five years, saved both to disk and in printed log reports, I offer what I argue are the essential characteristics of such transformative teacher networks—that is, I name and illustrate characteristics of the culture of a network that helps it become a "hothouse" for teacher professional development. Additionally, I offer a few observations on what I feel are the opportunities and constraints inherent in housing transformative teacher networks in a virtual space. (Swenson, 2003, p. 265)

Peter Smagorinsky et al. Lortie's (1975) observation that schools tend toward conservative practices remains true over a quarter-century later, as indicated by many studies of school practice (e.g., Borko & Eisenhart, 1992; Ritchie & Wilson, 1993; Smagorinsky, 1999). Meanwhile, as many conservative critics (Gross, 1999; Stotsky, 1999) have argued, new teachers are trained in schools of education that tend to espouse more liberal pedagogies. This chasm between university and school has often created a tension for education students who are immersed in a liberal culture during their university course work and must practice in conservative school environments. Our goal in this article is to focus on one teacher who experienced this tension. (Smagorinsky, Lakly, & Johnson, 2002, p. 188)

Janet Kaufman. After jogging up two long, wide steep flights of stairs to the fourth floor on a Thursday afternoon, I catch my breath and walk three doors down the hall to the Family Literacy Center. It's a big classroom that the principal has given over to us for the last four years—to me, my university preservice teacher students, and the high school students and teachers who attend and work at the school which houses the Center. The mural that my students began painting with the high schoolers four years ago has spread to all four walls, and the fluorescent lights overhead have been turned off in deference to low reading lamps we've brought in. A clothesline stretches from one wall to the other, with kids' poetry hanging from it. And on a small table stands a display of book art: houses and bodies with pages telling life stories of the students who made them. ... After four years, the Center has become what the kids describe as a "safe place." When we've asked them what the word "family" in Family Literacy Center means to them, they talk about how the people who direct the Center and those who come to if feel like family. But for me and my students, it has also become a place to investigate the role of teacher–student or mentor–student relationships in learning and literacy development. I have come to ask two primary questions: How does the service-learning project change the way my students learn, and how does it change the way I teach? (Kaufman, 2004, pp. 174–175)

Susan Wall. My narrative examples in this essay draw on teacher research produced by participants in the Institute on Writing, Reading, and Teaching at Martha's Vineyard, a graduate program sponsored by my English Department. ... The three teachers whose work I discuss here generously allowed me to interview them about their research experiences and portfolios after their projects were completed. Their portfolios included, in addition to their final research reports, their authors' field notes, drafts, and letters to and from [the professor of the course]. I am interested in the challenges they faced in writing their research accounts and in the ways they represented themselves in those narratives. My approach, in other words, is rhetorical rather than empirical or pedagogical. That is, I read particular teacher researchers' texts as portraits of writers writing in much the same way that literary critics use methods of rhetorical criticism to study first person accounts: by analyzing those texts and by taking into account what their authors say about how they came to be written and what was at stake in their creation. (Wall, 2004, pp. 290–291)

As we read these four passages, the first thing that strikes us is how differently the authors position themselves, even as they all use the first person (*I* or *we*). We would claim that all of them construct themselves as knowledgeable researchers, but that they do so in quite different ways. Swenson, for example, establishes herself as knowledgeable by her use of explicit and perhaps insider terminology ("on-line 'transformative teacher networks'" and "virtual space"), by references to a schema she created of "essential characteristics" to explain her research, and by her long investment in this research (5 years). Smagorinsky, Lakly, and Starr heavily cite the work of other researchers and thus situate themselves in terms of what others in the field have said, an indication that they are following a long line of other, respected thinkers. Kaufman, through her storylike beginning, immediately places the reader in the midst of her context and thus establishes her knowledge through her intimacy with the research site. Finally, Wall strikes an interesting balance: showing herself both as a serious researcher capable of doing this work (by her association with literary critics and her knowledge of rhetorical strategies) and as a gracious learner who is not posing a new theory but rather "interested in the challenges" the teachers faced.

We also notice how these authors position themselves in relation to their research question. Kaufman, for example, names her questions in an upfront manner ("I have come to ask two primary questions: How does the service-learning project change the way my students learn, and how does it change the way I teach?")—which seems to us a way of establishing these as true questions arising from the context of her work, questions that she honestly wondered about. In many ways, she establishes herself as a learner, raising the kinds of questions any of us might raise as readers. The other researchers align themselves with their research questions in different ways. Swenson, for example, does not directly name a question (although you can easily imagine one underlying her work), but rather establishes herself as someone who after 5 years of research really does have a thoughtful answer. For Smagorinsky and his colleagues, the research question is less one that arises from their immersion in their own practice (as it seems to be for Swenson and Kaufman), as it is one that applies what others have said through a look at one teacher.

All of these approaches—different as they are—are legitimate ways of positioning yourself as author and researcher. By looking through these passages and the many others we have read in our careers, we have identified several considerations to keep in mind as you begin thinking about how to position yourself in relation to your research and in relation to the larger field, whether it be composition studies, rhetoric, English education, or technical communication. The important point here is this: Through the choices you make, you create an authorial self; thus, it is very important to be both thoughtful and deliberate in these choices. The considerations we have identified follow:

- *The language you use:* The language you use will go a long way toward positioning you as an experienced member or initiate in the field (and there might be good reasons why you might choose one over the other). Be thoughtful, for example, about how you use certain specialized terms and ideas: Do you want to present yourself as an expert who uses the terminology of the field with ease and grace? This might be an important stance to adopt if you are writing to an audience of established experts in the field; by using the language associated with the field in a sophisticated manner, you establish your credibility with those potential readers. If you do go this route, you should make sure that you understand these terms and the underlying history and rationales behind them. Having this understanding will allow you to integrate the terms fluently into your writing. Alternatively, you might choose to position yourself as more of a learner and questioner in the field by using commonplace words and providing full explanations when you use more specialized terms. This might be the way to go if you want to place yourself on an equal footing with a particular audience, casting yourself as a learner along with them. Similarly, think about the kinds of sentences you construct. Long sentences with complex clauses establish an authorial self quite different from that of an author who uses simple sentence structures.

- *How you use others' research:* We all use the research of others as a way of establishing our own authority—to prove we have done our homework and are aware of an ongoing conversation in our field. But the ways in which we use others' research can vary, and these variations can influence how we establish ourselves as authors. Sometimes, for example, beginning researchers use others' research as a bedrock for their own work, relying on it as a schema for which the newer researcher's work serves as a further example. In other words, the senior researcher's work establishes the boundaries for the discussion or even the format; the younger researcher's work is useful in that it further clarifies the other researcher's approach. For example, if you were researching the topic of revision, you might focus your writing on the approaches to revision demonstrated by Barry Lane in his book, *After the End* (1992). You could use your students' revisions as further proof that Lane's approaches work.

In terms of positioning yourself as an author, this approach casts Lane as the expert and you as the learner. An alternative way of using others' research is to place your work at the center but refer to other sources as evidence supporting your ideas. In other words, using the previous example, you write up your findings about revision by establishing your own schema and your own ideas, but you quote Lane and others in support of your work.

• *The stance you adopt:* Sometimes as authors, we want to create a clear, reportorial tone, one in which we present our findings in a very factual way that establishes an answer that is not easily questioned. Oftentimes, researchers take this stance when they want to propose a specific solution or approach to a problem. Think, for example, about a situation in which your research might be used to select a new approach to carrying out a task in an organization or to select a new software tool. In this case, a reportorial approach might be the best choice. However, at other times, we may want to present our findings in all of their complexity: in a way that suggests that there is no single, clear-cut answer, but lots of question still surrounding the findings. An example here might be when you have done pilot research and are writing up the results as a step toward moving the research forward. Again, as the author of this sort of text, you have the ability and right to choose the stance you take.

Prompt 5: *Thinking About How Authors Position Themselves.*

Select a journal in your field and look at several articles published in it (look at a minimum of four articles in issues that span no more than 2 years). Try to characterize how the authors position themselves in their article. Find excerpts that support your judgments. Be prepared to share your findings with your classmates.

Prompt 6: *Determining Your Own Authorial Position.*

Think about your own work and how you are planning to present it. Use the labels and explanations in this section to describe the authorial position you plan to take in your writing. Try to explain, as well, why you think the authorial position you plan to take is appropriate, especially given your audience(s), your purpose, and the genre you have selected.

Determining Your Focus or Cut

At this point, hopefully you have identified an audience and purpose, selected a genre, and thought about how you want to position yourself as a researcher and author. The next step, of course, is to dive into your writing, using all the strategies and tools that work best for you as a writer as you plan, draft, revise, and edit your final piece. As you approach this stage, there are many additional choices you need to make. One of these is to choose the primary focus for the project. (Another name for this is the cut or angle for the project.) What we mean here is that we find it very useful to think carefully about the specific focus of our work and especially to consider that focus in light of the many other possible ways in which we might focus it. Let's start by looking at an example from Cathy's research:

> When I was preparing to write up my research on community organizing and teaching, I went through all the steps listed above: I knew that my audience would be practicing teachers, especially those who were concerned about how to change the current negative rhetoric about teaching English/language arts. I knew my purpose was to help educate them about the world of community organizing and how the strategies community organizers regularly use could help them organize the parents in their own schools. I knew I wanted to write a professional book, the kind that is sometimes used in a course and sometimes just read by practicing teachers (although I did do lots of conference papers, workshops and two articles in journals on the way to the book). I also thought hard about how I wanted to position myself as a researcher and author, and I decided on a stance and tone that showed me as a questioner and learner, someone who had thought a lot about the best ways to reach out to parents and the community but at the same time someone who was a complete neophyte in the world of community organizing.

> When I began writing, though, I faced another important choice—how to focus or determine my "cut" on the issue. In other words, what would be the most effective way to get across my purpose to my specific audience? I could, for example, choose the angle that the work of community organizers is worth studying in detail because their work should serve as a model for ours. Or I could choose a different angle: that teachers and community organizers are a lot alike and that teachers are already using a lot of community organizing tactics but don't always know it. I chose the second angle for a number of reasons: because it seemed a good way to help teachers feel that learning new information about community organizing was within their grasp and because it built upon the important work that teachers already do. In my mind, approaching my argument in this way would make it more convincing for this particular audience.

Your focus or cut, in other words, goes beyond just your purpose; it serves as a kind of lens through which you filter and focus your data in order to achieve your purpose for the audience you have in mind. It might be helpful at this point to think for a moment about how your own work might be focused in slightly different ways to reach your audience and achieve your purpose. Here are some more examples:

• A teacher we worked with had done his research on writing across the curriculum and writing to learn, with the intention of creating a handbook to help the teachers in his school integrate these activities more into their curriculum. He gathered research in three main ways: by surveying the teachers about how much writing they did, how it worked for them, and what impediments stood in the way; by observing teachers in his school as they integrated writing to learn activities; by reading about theories and practices of writing across the curriculum in journals, books, and conference presentations. In writing up this handbook, he identified three possible ways to focus it: (a) Standardized tests today require students to write in all subject areas. Becoming aware of writing to learn activities can help teachers help students do better on tests. (b) Many teachers identified the two main difficulties in integrating writing to learn activities: time and lack of expertise. Learning how to deal with both of those issues can make a teacher's life easier. (c) Experts around the country demonstrate how much writing improves when students are asked to write across their school day and not just in English class. Understanding the theory behind writing to learn can help teachers figure out how to use more writing in their classes.

• Our student who researched Nobel Prize lectures in chemistry was interested initially in examining the rhetorical features of these lectures in a very general sense. His data set included the lectures of three chemists who had, as a group, won the Nobel Prize in chemistry in 1996. As his work progressed, and as the amount of data he had grew, he began thinking about how best to focus his analyses and his written reports of the work. That the chemist's discourse was epideictic or ceremonial was very apparent, but it also was deliberative and persuasive, and both "cuts" revealed interesting information about the features of these lectures and about how they function in the sciences. Whereas he focused his thesis quite broadly, covering both the epideictic and deliberative functions of the lectures (he was unable at this point to narrow his focus), he realized soon that he needed a much sharper focus in order to publish the work. However, focusing on one or the other exclusively (the epideictic vs. the deliberative) weakened his presentation. As a result, he decided to keep the focus on both but to cut out a good bit of the detail from his initial write-up. This enabled him to establish a stronger focus.

Determining your focus is clearly connected to audience, purpose, and genre. In order to select your focus, you need, first, to think carefully about these rhetorical notions and, second, to consider what you might foreground and what you might relegate to background information in order to make your point convincingly. For example, depending on how you choose to focus your work, you will likely foreground certain ideas from the literature of your field while simply acknowledging or just briefly mentioning others. In other words, thinking about your focus also often involves thinking about the scholarly conversation of your field and where and how you wish to enter it. Our best advice for how to do this most effectively is to identify several alternative ways to focus your work before you decide on the final one.

Being Convincing/Writing Persuasively

As we mentioned earlier when we quoted Clifford Geertz, part of your job as a researcher—and author of your research—is to be convincing: to help readers feel as if they truly understand why you are making the claims you are and that your claims make sense. That's not to say that all readers will agree with your conclusions; however, your goal is to write so that the reader can at least understand why you see your conclusions in the way you do.

A big part of this is to argue your case well: to be organized, to provide sufficiently detailed evidence for your claims, to be internally consistent, to stay focused, and to draw conclusions from your data that make sense. The guidelines provided to reviewers of a journal for which Ann reviews reinforce these ideas and provide a good sense of what's expected, at least by academic journals, for an article to be judged convincing. The following list is quoted directly from the reviewer guidelines for the journal, *Technical Communication Quarterly*:

- Is the topic relevant and of interest to *TCQ*'s audience?
- Is the article logically organized? Does it have a clearly stated purpose and adequate introduction?
- Is the design of the study being reported sufficient to support the argument and conclusions offered?
- Has the text been adequately researched? Are the claims backed up by pertinent research and reference material? Are the claims original? Is the author familiar with the existing, relevant scholarship?
- Are there any gaps in the text? Are assumptions made that need explanations?
- Is the conclusion accurate?
- Does the title reflect the content?

Reviewers for *TCQ* are asked to consider if an article if well organized, if its purpose is stated clearly, if the design of the study is adequate, if the research is sufficient, and if the claims are supported by the research. They are also asked to look at how authors situate their claims in the existing literature and what contribution the work makes to this literature. They look for gaps and unsubstantiated assumptions, and they also are asked to make judgments about accuracy.

Lists such as this one can also be useful for checking your own writing. Of course, it's difficult for most of us to judge our own writing objectively; however, we can review it with these elements in mind, and we can also ask others to consider these elements when they review our writing. Also, while elements like these seem obvious, they can sometimes be hard to achieve. Writing up research requires attention to detail and a great deal of care. You have to be precise and accurate and you have to present ideas in a clear, consistent, and coherent manner.

A number of the choices you need to make as you write up your research can help you with being persuasive. For example, one of the decisions you will make concerns what format and/or style to use to present the information you have. In qualitative research, one very common presentation format is the *case study*. In other words, it is quite common for qualitative researchers to focus on a single case in writing up their research (e.g., a student, a worker, a classroom, or an event). As an example, if you have studied in your research the writing practices of adults in the workplace, then you might focus in your write-up on just one of the adults. If you researched how teachers learn about new ideas in writing and transfer those ideas to their teaching, then you might address a particular staff development meeting. Zeroing in on a person or event, the argument goes, allows the researcher and the reader to focus on the detail of a single experience—a kind of zoom lens approach that lets the minutia and nuances of the experience take center stage. Addressing one case in depth, these researchers say, helps us make sense of the larger issues of the research study.

If you do select this approach for presenting your findings, you should keep several things in mind. First, the case you select to present should represent your research accurately and completely. If you choose to present an anomalous case (e.g., the only student in the room who didn't respond to a new way of teaching revision), then you should present your reasons for doing so and provide sufficient context for the case. You should also be careful to not overgeneralize from a case. Although cases can help audiences make sense of parts of the research, they seldom are generalizable to "all students," "all workers," or "all events." As a researcher, it is important that you acknowledge your understanding of this point. Finally, the cases you present should always be real. As you review your findings, it may become tempting to combine a little bit of one case with a little bit of another. Such a composite, it may seem, does a better job of making the points you wish to make. However, presenting a composite as a real case is unethical. In such situations, you should consider a different format to present your data so that you accurately and fairly represent what occurred.

Another common format or style for presenting qualitative research is *narrative*. Many qualitative researchers find that telling the story of their research (e.g., telling the story of a classroom or workplace situation) is the best way to convey the richness and complexity of their findings. This presentation style can be very compelling because it allows readers to feel as if they are actually there, in the situation, with the researcher—the feeling of "being there" that Geertz talks about. A narrative style lends itself well to detail; it allows you to recreate your experience in a way that helps the reader really understand your findings. For example, a researcher writing about struggling male readers may write a narrative about the classroom that allows us to picture these readers as they find various ways to refuse to read books (e.g., by arguing with the teacher, putting their heads down on the desk, using their iPods to tune out the teacher, slamming their books on the table, or continually talking to their friends). Such a narrative can be very convincing to a reader who may respond, "Oh, yeah, I've got students like that too," and who may read even more closely to see what solutions the researcher offers.

Like case studies, narratives also require consideration of certain issues. For example, narratives can tempt researchers to embellish details in order to make the writing "sound" good. We are accustomed to thinking about narrative in relation to fictional genres, so it's important to keep in mind that a narrative style of research writing needs to be as accurate and precise as any other style for presenting research findings. Also, as a writer of a research narrative, you need to be aware of how you present your research participants and yourself. If you find that you are becoming the star of your own story, for example, you might want to step back and tweak your focus. You also want to be sure to represent your participants fairly and accurately, which is something we say more about in the ethics section later. Typically, a qualitative research narrative should be about a situation, an event, a place, or people.

A third common format for research writing is the traditional *academic research paper* or report. This type of writing traditionally adheres to a fairly rigid organization: research question, literature review, methodology, findings, and then discussion and/or conclusions. However, different academic journals sometimes have different expectations or requirements, so it is important to look into this in advance, especially if you already know the journal you are targeting. This format generally calls for a style that's more reportorial—authors usually present claims and then provide support for those claims from their data (the authorial persona for this format can sometimes even seem distant). However, this format certainly does not rule out the presentation of a case or cases or the use of narrative. It is simply a more structured and regularized format that is quite common in academic scholarship.

We would like to make a few final comments about research formats as a natural extension of this discussion. The formats and styles presented here are just a few examples of those you might select. We have not tried to be exhaustive. Also, as some of our earlier comments suggest, none of these formats is mutually exclusive. In other words, you may well write a more traditional research report that focuses on a particular case from your research, or you might incorporate a narrative style into a case or into your research report. We separated these formats simply to address the options you have for making your research report convincing to your readers. The best way for you to choose a format is to think carefully about your audience and purpose as well as about your comfort level with these approaches. In order to get more comfortable with these formats (or at least to expand your repertoire of possibilities), we suggest taking time to read and explore different genres of research.

Prompt 7: *Selecting a Format and Style for Your Write-Up.*

Find examples of research write-ups that focus primarily on a case, that use narrative, and that are written in a traditional research report format. Keep in mind that these categories are not discrete (e.g., the example of narrative you find might also present a case) but simply ways to focus one's writing to achieve certain goals. Now think about your own research write-up and decide what format and style would be most effective given your own audience(s), purpose, and goals.

Obtaining and Using Feedback

All of us have probably encountered the image of the lone author writing in solitude, but few of us write completely in isolation. Writing, in fact, is a social act, even when we are writing by ourselves. So is research. From the start of your research, you have likely involved a number of individuals in your project. You have talked with them about your plans, gotten feedback from them on your question, shared your research proposal with them, gotten their ideas about methodology, and so on. Now that you are writing up your findings, you will also want their input. You can start by discussing with your advisor, or your peers, who your audience should be, what your purpose is, what genre you should use, what authorial stance you should take, and what style you should use. In other words, all of the decisions we have been discussing are those you will likely make with at least some input from others. As mentioned in chapter 6, you might even seek input from your research participants, which we certainly encourage you to do.

Both of us believe strongly that writing is inherently social and that writers benefit greatly from interactions with others during every stage of the writing process—planning and prewriting, drafting, and revising. In other words, we encourage you to seek feedback not just after composing an initial or final draft, but at every step along the way. And, if you have the opportunity, work closely with your advisor on your writing. Your advisor can help you shape your writing in ways that are consistent with the conventions and expectations of the field. Your advisor can also help you with voice, tone, style, and the arguments you craft, and with moving your writing toward publication.

You should also seek feedback from your peers—your classmates, others in the field, and even spouses and significant others. Some of our toughest critics and most careful readers over the years have been our own peers, including our spouses. Whether we simply converse with them informally about our plans for our writing or ask them to read a draft, we know we can count on their insights to improve our writing. Some writers even go a step further and join a writing group. The Six Scholars, the students whom we mentioned in earlier chapters, have stayed in touch for over 8 years, and they continue to support one another and read each other's writing. Similarly, a group of our literature colleagues have been reading each other's work and meeting regularly for a number of years.

These are both instances of groups that formed with individuals who all knew one another. You can also join a writing group with individuals you don't know. This can be beneficial because these readers can bring a fresh and unbiased perspective to your writing. Finally, you may be most comfortable with having a single individual read and respond to your writing; someone with whom you are comfortable and who you trust as a reader. For this book, for example, we asked Rhonda

Copeland, a former student, to read and edit our drafts. We did this because of our familiarity with and trust in Rhonda's editing skills.

Sometimes an individual's or group's effect on a text seems so substantial that the author wonders if it is really more of a collaboration. This happened once to a group of scientists with whom Ann worked—a reviewer made such substantive contributions to their text that the authors added this individual as an author. In most cases, however, it's sufficient simply to acknowledge another individual's help. We see this frequently in acknowledgment sections for books, as well as in acknowledgment notes with journal articles. What's most important is to value the feedback you receive, to express your gratitude to your reviewers publicly and privately, and, if appropriate, to return the favor.

In short, we encourage you to seek feedback at every stage of your writing, from the moment you begin thinking about audience and purpose to the point at which you are ready to submit your manuscript for publication. Your needs at these different stages will vary, and you should remember to communicate those to the individuals from whom you are seeking advice. For example, if you have done as much work as you think you can at a certain point, and you feel confident that the manuscript is ready to be submitted, then you may ask a reader simply to copyedit. On the other hand, if you are just beginning to draft, you will probably want to ask a reader to consider larger issues such as meaning, organization, logic, flow, coherence, and voice. A reader's feedback certainly can shape the meaning that you convey with your text and even your interpretations of your findings. These individuals can play an important role in and contribute in significant ways to your writing.

Prompt 8: *Obtaining and Using Feedback.*

Think for a moment about the various ways in which individuals with whom you have interacted during your research have influenced it and contributed to what you have done. Next, begin thinking about how these individuals, and/or others, might contribute to your writing. Develop an informal plan for obtaining feedback on your writing. Identify the stages at which you might want feedback, the type of feedback you would like at those stages, and from whom you would ideally like to receive the feedback. Think also about how you will solicit the feedback (how you will approach and ask others to review your work; how you will locate a writing group, if that's your preference), and about how you will acknowledge and express gratitude to the individuals who give you feedback.

Letting Your Writing Sit

Another productive strategy that we often suggest to our students, although we sometimes encounter resistance to it, is to put your writing aside for awhile, especially when you are having difficulty with it. The saying that time cures all ills can be especially true with writing. Both of us certainly have had the experience of "getting stuck" with our writing and then putting it aside for a few days (even weeks) only to return to it with a fresh perspective. It's easy to get into a situation where you just can't seem to get anywhere with your writing, especially if you are feeling pressured or rushed with it. Or perhaps you have written a draft and received some harsh feedback. The feedback might be so hurtful (for most of us, writing is also very personal) that you end up needing time to regain your perspective.

Regaining perspective is really what this strategy is about. Writing is intense. Our own experiences reinforce that. For example, as Ann was writing the first draft of this section of the chapter, she was thinking about the pressure she was under to finish the chapter so she could get the manuscript, which was otherwise complete, to the publisher. The kinds of pressure we sometimes feel when we write, whether real or perceived, can easily lead to writer's block. It can also curtail our writing, causing us to think about certain issues or ideas and not others as we write.

There are a number of strategies that can help in these situations. Sometimes it's enough just to take a short break from our writing—to take a walk, run an errand, work out, or even take a nap. Sometimes it's more productive to put aside our writing for an afternoon or evening and start fresh in the morning. Other times, we need to put it aside for a longer period—a couple of days, a week, or even longer. Some refer to these rest periods as incubation, a term we like because it suggests that we can still be thinking about our writing. In other words, as Ann likes to tell her students, a writer's mind is always writing. So even if we put our writing aside for a time, we can still be thinking about it and generating ideas for it. We probably all have had the experience of thinking of things as we lay in bed at night, or as we take walks or run errands. Again, always having something handy to record one's thoughts, as we have mentioned in other places, can be useful in these situations.

Ultimately, you have to be the one who determines when you need a break from your writing, although some of us need to train ourselves to recognize these times and to give into them. Some writers even find it helpful to plan breaks deliberately. For example, one or our students told us that she intentionally scheduled two week-long vacations during the time in which she would be writing her master's project. She felt she needed something to look forward to, and a deadline, to keep her motivated. She also knew she would need the breaks, and she was disciplined enough to make it work (a common hazard in cases like these is that the writer procrastinates and doesn't accomplish anything by the first break).

Writers ultimately need to know themselves to make the best determinations with respect to taking breaks from their waiting. Your personal style, your circumstances, and your own level of comfort with writing will all figure into your approach. For example, if you have a short amount of time to write up your research (e.g., because of a semester deadline), you probably won't have the luxury of taking many or very long breaks from it. If you are less pressured, then you may be able to take as many or as long of breaks as you need. Whether long or short, or few or many, the key benefit from taking breaks is the critical distance they afford. There is nothing like having a fresh perspective on our writing to sustain it and make it better. What's most important is to know that occasional, and sometimes even frequent, breaks from your writing can be a productive writing strategy.

WRITING THROUGH VARIOUS LENSES

Throughout this book, we have talked about the role of lenses during every stage of research. Their role is probably most obvious at the interpretation and writing stages. Our lenses determine how we see our data and the meaning we ultimately derive from and attach to it. They also are a determining factor in how we convey that meaning to others. In other words, our lenses influence not only how we end up seeing our research, but also how our audiences end up seeing it: what significance they attach to it, what meaning they derive from it, what contribution they see it making. Different lenses—for example, different ways of situating our work and different theoretical perspectives—can alter our meaning and the contribution the work makes. For example, Ann used situated learning theory (Lave & Wenger, 1991) as her primary lens for examining the mentoring relationship that existed between a physicist and his graduate student. She showed how the relationship resembled that between a master and apprentice and how the student acquired knowledge of the rhetorical conventions of the discipline through a gradual process in which he began using those conventions. She also showed how the student was closely supervised by his advisor during that process. In other words, Ann showed how what the student experienced was a kind of legitimate peripheral participation, a concept Lave and Wenger address. Her research, and the meaning she constructed from it, would have been much different if she had selected a different lens (e.g., she might have adopted a more language-oriented lens and focused less on the teaching and learning that occurred in the situation and more on the conversational dynamic and how that reinforced the authority relationship between the advisor and his student). Her readers would have taken something very different from the latter focus than they did from the former.

Margery Wolf's (1992) work, discussed earlier, offers another good example of how lenses influence our writing. Wolf not only used different genres to convey the same findings, she also used different interpretive lenses. The three versions of her research convey very different meanings and perspectives on her work. She stresses that writing is a rhetorical act and that interpretation, which depends on lenses, is a key part of our research and writing. When you write, you choose which events from your research to foreground and what themes to highlight. The lenses you select influence not only your own understanding of your findings but that of your readers as well. They also can influence the style, format, and genre of our research—the various tools and choices we have addressed throughout this chapter.

What is most important to remember about writing in this sense is that it is, as Wolf stresses, always a rhetorical act. We have the inherent advantage as members of our field of grasping what that really means. There are many choices involved in writing, and lenses are some of the most significant of these, particularly given their implications for the meaning we convey. We should make these choices as consciously as we can and also make them explicit to our readers. When warranted, we might even want to explain to our readers why we made them. We also need to understand that not all of our readers, or even our participants, will agree with the lenses we select, or how we use them. Borland's work, addressed in chapter 6, offers a good example of this. Borland's primary participant, her grandmother, took issue with Borland's interpretation of the events she observed. Her grandmother especially took issue with the feminist lens Borland used in her interpretations, which the grandmother said held little meaning for her. If your participants will be reading your work, then it is important to remember that they may not agree with or fully understand the lenses you use. How you respond to your participants if they do disagree is something you certainly will want to consider.

Journal reviewers may also question our lenses. Reviewers are asked to comment on how authors situate their work: whether the literature that's presented is the most appropriate, whether it's interpreted and conveyed accurately, and so forth. And they usually provide such feedback. In one instance we know of, a reviewer told the author that she had completely misread the work of another author whom she had cited heavily. Because the author was citing her own previous work (the review was blind), you can imagine her surprise at hearing that she was misinterpreting herself! In most cases, however, reviewers raise legitimate questions about authors' lenses and about how they situate their work. Sometimes they suggest works the authors haven't considered, or they suggest different interpretations or even different lenses (e.g., different ways to frame the work theoretically). What is most important to remember is that lenses, like so much else in our research, are dynamic and certainly not binding. Our choices of and sense of our lenses can benefit as much from others' feedback as the style and format of our writing. In short, meaning, and, connected to that, lenses, are always negotiated between authors and readers.

WRITING ETHICALLY

Because writing makes our research public, ethics are again a significant concern at this stage in the research process. Qualitative researchers need to think about how

Prompt 9: *Identifying and Making Your Lenses Explicit in Your Writing.*

Think about and identify the primary lens you are using in your work. Think of one or two alternative lenses that you either considered or that might also be appropriate. Freewrite about how these might change your work, especially how you write it up and present it to your audience(s). Also write about any concerns you have about your lenses (e.g., whether your participants will agree with your interpretations and how you frame them, whether you have addressed all of the relevant literature, etc.). Depending on how far along you are with your writing, give a draft of your work to a peer and ask that person to focus on and respond to how you situate and frame your findings.

---

they present (and represent) their participants. They need to guard against taking their findings out of context, embellishing them, conveying information that participants have asked them not to, taking liberties to make particular points, excluding information that should be presented, or drawing conclusions that their data won't support. As qualitative researchers, our responsibility is, first and foremost, to our participants. The writing stage brings this responsibility to the fore again. We are responsible both for the integrity of our findings and for the integrity of the texts we generate with those findings.

The issue of responsibility to our participants warrants further discussion. Different researchers interpret this responsibility in different ways, and the ways you choose will be entirely up to you. At one end of the continuum, some researchers seek to co-construct meaning with their participants and to make them true collaborators in the research, at times even asking them to be coauthors (the latter, in practice, is still pretty rare). On the other end of the continuum, some researchers uphold their own expertise and limit the roles their participants play (participants are sources of data to these researchers). Of course, most researchers fall somewhere in the middle—involving participants in various ways at different stages of the research. For example, some of us like to have participants, if they are willing, read and respond to drafts of our work. Some researchers like to discuss their interpretations with participants even before they begin writing. Like so much else with qualitative research, what you prefer will depend on the situation and on your stance and ethos as a researcher.

In the very least, you need to be accountable to your participants—you need to keep any promises you make (e.g., to not use certain information), you need to be cognizant of how your writing portrays your participants (e.g., what if your findings end up portraying them in a less than favorable light?), and you need to grapple with any dilemmas that arise in your relationships with them (e.g., whether to include information simply because they want you to include it). All of this underscores why it is so important to be self-conscious and reflective as researchers and to be explicit about our perspectives on our research and the lenses and biases we bring to it. We repeat here advice that we shared earlier: Balance your own expertise with the expertise of your participants, listen attentively to what your participants tell you, question your interpretations constantly, and realize that *truth* is a pretty elusive concept.

Another important ethical issue in writing concerns how we use and acknowledge others' ideas. We can all think of instances of writers being accused of plagiarism, but there are other ways to misuse or misrepresent the ideas of others. As writers, we need to be cognizant of how we use the words of our participants and our sources. It isn't difficult to cite sources properly, and we should be diligent about doing so. There are also many tools available now (e.g., RefWorks and EndNote) that provide assistance. Having others, including our participants, read our work is one way to check its integrity. We can ask our readers to pay attention to how we use sources, how we present our findings, and how we portray our participants.

Ultimately, our interpretations must be supported by the data we present, and that data should be sufficient. One common criticism of qualitative research is that some researchers make claims beyond the scope of their research. Because most qualitative research in our field focuses on limited situations or cases, making broad generalizations based on those situations is not appropriate. This does not mean that qualitative research is any less useful or that it does not contribute anything. What it does mean is simply that we need to provide a proper context for our work and be honest about its scope—we need to hedge appropriately. Data that represents specific instances can be very informative and can help us expand our understandings of issues in our field. It is valuable, for example, to take a particular research case and place it up against many research cases in order to begin seeing similarities and contradictions at work. So qualitative research, even when focused on particulars, still makes important contributions to our professional knowledge, but we need to be honest, both with ourselves and with our readers, about what those contributions are.

Prompt 10: *Thinking About Ethics.*

Identify at least two ethical issues you have encountered while carrying out your research and reflect on your responses to them. Next, think about the ethical concerns you have as you approach the writing stage of your research (e.g., representing the voices of your participants accurately and fairly; hedging appropriately, but not too much; making claims that are supported sufficiently by your findings and that are appropriate given the scope of your research). Talk through your concerns with a friend, preferably someone who will also be reading drafts of your work. Think about and discuss with this individual how you might attend to these issues as you write.

CONCLUSIONS

Writing certainly can be one of the most challenging tasks you undertake as a researcher, but it can also be the most rewarding. There is nothing quite as satisfying professionally as sending off a finished, polished draft for publication; receiving an acceptance notice from a journal; or publishing your first book. And depending on how far you have progressed in your research, you may be well on your way to achieving those milestones.

Throughout this book, we have tried to lay out a process for undertaking qualitative research that will help you become a careful and successful researcher. We hope that you take from this book a sense of qualitative research as a dynamic and recursive enterprise that can contribute to our understandings of a vast array of issues. Whether you research for personal or professional reasons and, if the latter, whether you use the results yourself or share them with others, you can hopefully feel confident now that you have carried out your research thoughtfully, systematically, and ethically. We hope you can also assert and feel confident about your identity as a writing researcher. If you have read through all of the chapters of this text and completed even some of the stages of your research, then you have made significant strides in acquiring this identity. You have, in fact, become a writing researcher.

References

Adler, P. A., & Adler, P. (1994). Observational techniques. In N. K. Denzin & Y. S. Lincoln (Eds.), *Handbook of qualitative research* (pp. 377–392). Thousand Oaks, CA: Sage.

Anderson, K., & Jack, D. C. (1991). Learning to listen: Interview techniques and analyses. In S. B. Gluck & D. Patai (Eds.), *Women's words: The feminist practice of oral history* (pp. 11–26). New York: Routledge.

Anderson, P. V. (1996). Ethics, institutional review boards, and the involvement of human participants in composition research. In P. Mortensen & G. E. Kirsch (Eds.), *Ethics and representation in qualitative studies of literacy* (pp. 260–285). Urbana: NCTE.

Anderson, P. V. (1998). Simple gifts: Ethical issues in the conduct of person-based composition research. *College Composition and Communication, 49*, 63–89.

Barton, E. (2000). More methodological matters: Against negative argumentation. *College Composition and Communication, 51*, 399–416.

Berthoff, A. E. (1987). The teacher as researcher. In D. Goswami & P. Stillman (Eds.), *Reclaiming the classroom: Teacher research as an agency for change* (pp. 28–39). Upper Montclair, NJ: Boynton/Cook.

Blakeslee, A. M., Cole, C. M., & Conefrey, T. (1996). Constructing voices in writing research: Developing participatory approaches to situated inquiry. In P. Mortensen & G. E. Kirsch (Eds.), *Ethics and representation in qualitative studies of literacy* (pp. 134–154). Urbana, IL: NCTE.

Bloome, D., Shuart-Faris, N., Power Carter, S., Morton Christian, B., & Otto, S. (2005). *Discourse analysis and the study of classroom language and literacy events: A microethnographic perspective*. Mahwah, NJ: Lawrence Erlbaum Associates.

Bomer, R., & Bomer, K. (2001). *For a better world: Reading and writing for social action*. Portsmouth, NH: Heinemann.

Borland, K. (1991). "That's not what I said": Interpretive conflict in oral narrative research. In S. B. Gluck & D. Patai (Eds.), *Women's words: The feminist practice of oral history* (pp. 63–75). New York: Routledge.

Brandt, D. (1994). Remembering writing, remembering reading. *College Composition and Communication, 45*, 459–479.

Brandt, D. (1995). Accumulating literacy: Writing and learning to write in the twentieth century. *College English, 57*, 649–668.

Brandt, D. (2001). *Literacy in American Lives*. Cambridge: Cambridge University Press.

Brown, G., & Yule, G. (1983). *Discourse analysis*. London: Cambridge University Press.

Burke, K. (1968). *Language as symbolic action: Essays on life, literature and method*. Berkeley: University of California Press.

Calkins, L. (1983). *Lessons from a child: On the teaching and learning of writing.* Portsmouth, NH: Heinemann.

Cameron, J. (1992). *The artist's way: A spiritual path to higher creativity.* New York: Jeremy P. Tarcher/Putnam.

Charney, D. (1996). Empiricism is not a four-letter word. *College Composition and Communication, 47,* 567–593.

Charney, D. (1998). From logocentrism to ethnocentrism: Historicizing critiques of writing research. *Technical Communication Quarterly, 7,* 9–32.

Emerson, R. M., Fretz, R. I., & Shaw, L. L. (1995). *Writing ethnographic field notes.* Chicago: University of Chicago Press.

Finders, M. (1997). *Just girls: Hidden literacies and life in junior high.* New York: Teachers College Press.

Finley, A. (2005). The interview: From neutral stance to political involvement. In N. K. Denzin & Y. S. Lincoln (Eds.), *Handbook of qualitative research* (pp. 695–728). Thousand Oaks, CA: Sage.

Florio-Ruane, S., & Morrell, E. (2004). Discourse analysis: Conversation. In N. Duke & M. Mallette (Eds.), *Literacy research methodologies* (pp. 42–61). New York: Guilford.

Fontana, A., & Frey, J. H. (1994). Interviewing: The art of science. In N. K. Denzin & Y. S. Lincoln (Eds.), *Handbook of qualitative research* (pp. 361–376). Thousand Oaks, CA: Sage.

Fontana, A., & Frey, J. H. (2005). Recontextualizing observation: Ethnography, pedagogy, and the prospects for a progressive political agenda. In N. K. Denzin & Y. S. Lincoln (Eds.), *Handbook of qualitative research* (pp. 729–746). Thousand Oaks, CA: Sage.

Frank, C. (1999). *Ethnographic eyes: A teacher's guide to classroom observation.* Portsmouth, NH: Heinemann.

Geertz, C. (1983). *Local knowledge: Further essays in interpretive anthropology.* New York: Basic Books.

Geertz, C. (1988). *Works and lives: The anthropologist as author.* Stanford, CA: Stanford University Press.

Gilligan, C. (1982). *In a different voice: Psychological theory and women's development.* Cambridge, MA: Harvard University Press.

Glesne, C. (1999). *Becoming qualitative researchers: An introduction.* New York: Longman.

Gluck, S. B., & Patai, D. (Eds.). (1991). *Women's words: The feminist practice of oral history.* New York: Routledge.

Goldman, S., & Wiley, J. (2004). Discourse analysis: Written text. In N. Duke & M. Mallette (Eds.), *Literacy research methodologies* (pp. 62–91). New York: Guilford.

Gurak, L. J., & Silker, C. M. (1997). Technical communication research: From traditional to virtual. *Technical Communication Quarterly, 6,* 403–418.

Halliday, M. A. K., & Hassan, R. (1976). *Cohesion in English.* London: Longman.

Hodder, I. (1994). The interpretation of documents and material culture. In N. K. Denzin & Y. S. Lincoln (Eds.), *Handbook of qualitative research* (pp. 393–402). Thousand Oaks, CA: Sage.

Hubbard, R., & Power, B. (1993). *The art of classroom inquiry.* Portsmouth, NH: Heinemann.

Hubbard, R., & Power, B. (1999). *Living the questions: A guide for teacher-researchers.* York, ME: Stenhouse.

Huckin, T. N. (1992). Context-sensitive text analysis. In G. Kirsch & P. A. Sullivan (Eds.), *Methods and methodology in composition research* (pp. 84–104). Carbondale: Southern Illinois University Press.

Kaufman, J. (2004). Language, inquiry, and the heart of learning: Reflection in an English methods course. *English Education, 36,* 174–191.

Kirsch, G. E. (1997). Multi-vocal texts and interpretive responsibility. *College English, 59,* 191–202.

Kirsch, G. E., & Ritchie, J. S. (1995). Beyond the personal: Theorizing a politics of location in composition research. *College Composition and Communication, 46,* 7–29.

Lane, B. (1992). After THE END: *Teaching and learning creative revision.* Portsmouth, NH: Heinemenn.

Lave, J., & Wenger, E. (1991). *Situated learning: Legitimate peripheral participation.* Cambridge, England: Cambridge University Press.

MacLean, M., & Mohr, M. (1999). *Teacher researchers at work.* Berkeley, CA: National Writing Project.

Moss, B. J. (1992). Ethnography and composition: Studying language at home. In G. Kirsch & P. A. Sullivan (Eds.), *Methods and methodology in composition research* (pp. 153–171). Carbondale: Southern Illinois University Press.

Odell, L., Goswami, D., & Herrington, A. (1983). The discourse-based interview: A procedure for exploring the tacit knowledge of writers in non-academic settings. In D. Mosenthal, L. Tamor, & S. Walmsley (Eds.), *Research on writing: Principles and methods* (pp. 221–236). New York: Longman.

Perakyla, A. (2005). Focus groups: Strategic articulations of pedagogy, politics, and inquiry. In N. K. Denzin & Y. S. Lincoln (Eds.), *Handbook of qualitative research* (pp. 887–911). Thousand Oaks, CA: Sage.

Power, B. (1996). *Taking note.* Portland, ME: Stenhouse.

Ruddock, J., & Hopkins, D. (Eds.). (1985). *Research as a basis for teaching: Readings from the work of Lawrence Stenhouse.* London: Heinemann.

Scheurich, J. J., & McKenzie, K. B. (2005). Analyzing talk and text. In N. K. Denzin & Y. S. Lincoln (Eds.), *Handbook of qualitative research* (pp. 869–886). Thousand Oaks, CA: Sage.

Seidman, I. (1998). *Interviewing as qualitative research: A guide for researchers in education and the social sciences* (2nd ed.). New York: Teachers College Press.

Sipe, R. B. (2003). *They still can't spell? Understanding and supporting challenged spellers in middle and high school.* Portsmouth, NH: Heinemann.

Smagorinsky, P., Lakly, A., & Johnson, T. (2002). Acquiescence, accommodation, and resistance in learning to teach within a prescribed curriculum. *English Education, 34,* 187–213.

Swenson, J. (2003). Transformative teacher networks, on-line professional development, and the write for your life project. *English Education, 35,* 262–321.

Wall, S. (2004). Writing the "self" in teacher research: The potential powers of a new professional discourse. *English Education, 36,* 289–317.

Wise, R. (1995). *Learning from strangers: The art and method of qualitative interview studies.* New York: The Free Press.

Wolf, M. (1992). *A thrice told tale: Feminism, postmodernism and ethnographic responsibility.* Stanford, CA: Stanford University Press.

Yagelski, R. (2001). It's (my) story, but I'm (not) sticking to it. *College Composition & Communication, 52,* 655–658.

Author Index

Subject Index